How To Build A Repro Rod

By John Thawley

Editor Steve Smith
Associate Editor Georgiann Smith

ISBN #O-936834-34-X

Published by

P.O. Box 11631/Santa Ana, CA 92711/(714) 639-7681

Dedication

This book is dedicated to may wife Terri who tolerated the project with a degree of humor; to my daughters Lori and Jenny who thought at various times my roadster was cute, funny looking, weird, and should have been painted a different color; to my son John T., who thinks the world would be a lot better if we all ran blowers and alcohol on the street; to my banker, who never developed a sense of humor concerning the project; and to suppliers and rod builders across the nation responding to my requests for photos and information. Sincere thanks to everyone.

Among those who should be singled out for special recognition are: Magoo, Warren Boughn, Paul Bonant, Lance Troupe, Brian Brennan, Joe Mayall, Ron Francis, Judy and Jim Kirby, Dick Williams, Pete & Jake, Jerry Kugel, Jim Ewing, Gary Lick, Bill Paul, Ed Moss, BF Goodrich, Herb Fishel, Terry Clark, Ed Hajek, Neil Gates, Roy Fjasted, Frank Aldana, Classic Instruments, Posies, Bob Lindebaum, Roger Schumacher and Sam McFarland.

Foreward

Tex Smith said when the time came he would write the foreword to this book. The time is here and the Smith foreword is not. This simply means Tex is out messing around in the garage and if there is to be a foreword I will have to write it. Here goes.

In writing this I feel as if I'm being prodded to confession by a shrill voiced female. It is to no avail. I have been a hot rodder-tinkerer, bench racer, garage messer upper, race car breaker, thread-stripping scroungy-dresser for more than thirty years. Mules pull pretty good but are difficult to turn.

Umpteen cars, engines, dollars, parts and race track hot dogs later, I have discovered the true salvation for all of us afflicted by the same symptoms — build a Repro Rod. Street Rodding is an individual expression in personal transportation; it is also an excellent opportunity to goof off and have a lot of fun. So . . . have fun.

John Thawley
Author/Tinkerer

TABLE OF CONTENTS

Introduction

This book is bound to rattle some cages. It was intended to do so. There are about as many ways of building a street rod as there are to write a book. This book is about how to build a "repro rod" — a reproduction of a hot rod. We've chosen the repro rods as the subject for a book simply because their time has come.

There will be those who say repro rods are not "real hot rods"; that they are store-bought toys requiring no skill to assemble. There will be sneers and slurs that all it takes is plastic money to build a plastic car. So be it.

Reproduction hardware means it is a lot easier to complete a car. Reproduction hardware means you don't have to be a body man, a welder, a machinist, etc., to complete a neat, safe car.

Reproduction hardware means you don't have to work on a rod for two to five years in your spare time in order to complete a car. Reproduction hardware means a lot more rod enthusiasts can now be active participants in the fun of street rodding.

This book does not concern itself with hammer welding, metal shrinking or the intracacies of chopping, channeling, boring and stroking. We will attempt to show how a safe, legal street rod can be assembled in a minimum amount of time for a far lower level of skills needed to do the same thing with wrecking yard and swap meet parts.

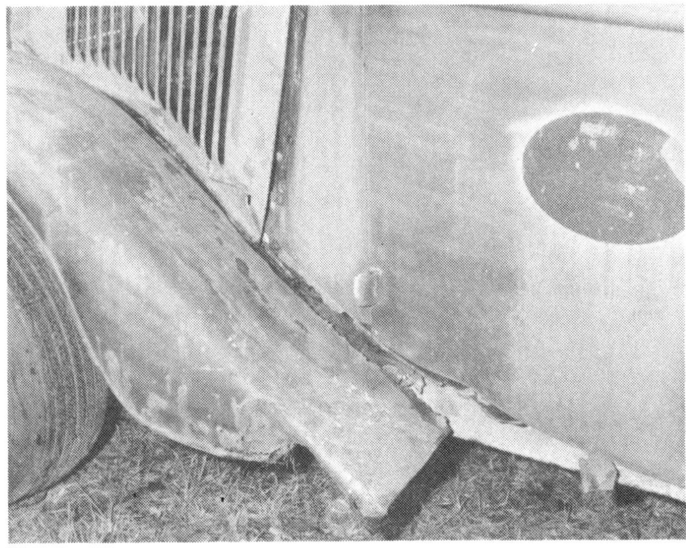

Sheet metal in this condition triggered a fiberglass replacement parts business which in turn launched the repro rod industry.

Your repro rod can be constructed to reflect your tastes in distinctive motoring pleasure. You'll be time and money ahead on the project by figuring out exactly how you want the car to look before starting to buy parts.

If you think building a repro rod is a lot of work, you're right, but when compared to bringing an original back to life, it's a snap.

The Plan of Attack

Because you are reading this book, it is reasonable to assume you have some interest in owning a "rod." The arena of rodding is now so large in terms of the types of cars involved that one has best be very specific about what he has in mind when the word "rod" is mentioned. To one, a '55 Chevy is a rod, to another it is a street machine. There are some who will never accept any car with fenders and a top as a rod. Others must have a supercharger or at least multiple carburetion as a criteria for a rod. Some will accept a closed car with a stock engine and a non-stock paint job and a set of after-market wheels as all that is needed to have a rod.

Thus, it is only fair that you know what we have in mind when we use the phase "repro rod". To our way of thinking, a repro rod is a car constucted solely or primarily with reproduction hardware. Due to the availability of hardware, the car will appear to be a Ford T, A or '32, '33, '34, or '36. It can be open, closed and may or may not have fenders. It can have most any running gear and suspension which is available, complete or in kit form.

In conjunction with putting this book together, the repro rod which is pictured throughout this book was assembled. We do not consider this car the ultimate hot rod, or the ultimate or only way to build a repro rod. The car shown here is simply one example of how a repro rod can be put together — with emphasis on problems and solutions.

Above all else, a repro rod should not be confused with a "kit car". To us, a kit car is a fiberglass body which is used to replace the body of an otherwise stock vehicle. Over the past twenty years there have been numerous designs and names — most of which were intended to be used on Volkswagen floor pans. Kit cars have primarily been distinguished by their sales pitch along the lines that to get one running all you needed was a weekend and a couple of buddies to help you lift one body off and another one onto your existing running gear. The pitch continued that you needed no special tools or skills, and each body came complete with a list of "easily obtainable parts" that could be purchased at the local wrecking yard for less than a hundred dollars. The parts list then went on to list the windshield from a '54 Stude coupe and the door glass mechanism from a '51 Ford pickup.

Pete Chapouris of Pete and Jake's summed up the kit car perfectly when he said "there is no light at the end of the tunnel." Simply stated, the kit car is far more than the home builder could handle.

The repro rod should not be confused with building or trying to build a rod from old parts. Jerry Kugel of Kugel's Kustom Komponents said he figured that "95% of the hobbiest cars started are never finished by the guy that starts them." The cars go from one garage to another and

Starting with the frame, then the running gear, begin assembling the car.

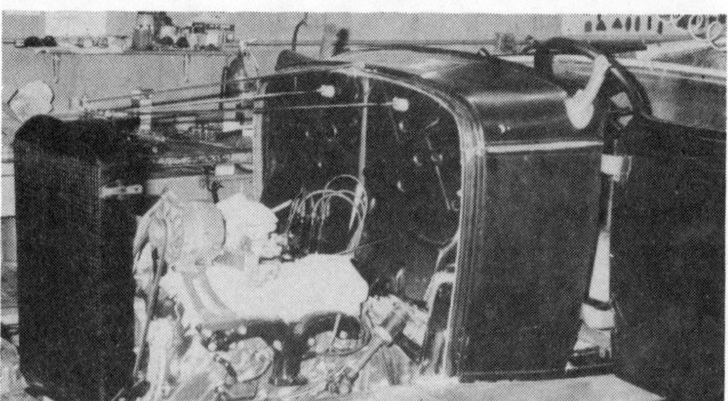

With running gear in place, put the body in position so details can be worked out on such items as battery box, gas tank, steering column, brake pedal and so on.

in some instances surface yearly at swap meets. We personally know of a '32 roadster that has gone through five owners in ten years and has yet to turn a wheel under its own power. The car has variously been fitted for an early Chrysler hemi, a small block Chevy, and now a small block Ford. None of these engines has ever been fired in the '32 frame. Over the years hundreds of dollars have been spent on hardware for the car — some new, some old, some reproduction, and still the car does not run. At least four of the previous owners have lost their once burning interest to own a rod.

Why don't cars get completed?

Pete Chapouris: "There are very few people who are capable of building a complete rod from scratch. This is supposed to be fun, relaxing, a hobby; and everywhere the builder turns he is butting his head against the wall. This won't fit, that has to be machined, that has to be welded and this piece needs to be bent. The would-be rodder buys a part from a manufacturer six states away and takes it to the local welding shop to get it installed. The welder does not communicate well with the rodder and the rodder only glances at the instructions and turns the project over to the welder. The nightmares increase."

Are we trying to talk everyone out of building a rod, or are we trying to say all problems magically disappear when building a repro rod? Neither. What this book will attempt to show is that with a plan of attack in using repro rod hardware there can be plenty of light at the end of the tunnel and that a repro rod *really can* be built.

If you are a dentist and you are into water skiing, you would not think of building your own skis or building your boat; if you are a real estate salesman and you are into bicycling chances are you would never consider building your own bike. But until recently, rodding was a cat of a different stripe. If you wanted a rod you bought an existing piece or tried to do it yourself. The latter is the

95% that Kugel mentioned.

But now the picture has changed. There is repro rod hardware available which greatly lowers the level of skill needed to complete a car. A fiberglass body means you don't have to know how to beat out dents and patch rust spots. Repro frames mean you don't have to know how to box a set of rails or weld up a broken crossmember. There are problems to be sure, but all too many of them are created by a misunderstanding (or no understanding) on the part of the rod builder as to what it takes to get the job done.

If you are a dentist you can undoubtedly make more money at what you do than you can welding engine mounts to a frame or trying to figure out some suspension problem. You have the skills and the tools to do what you do best. You have minimum skills and tools in the area of rod building, so why not pay to have someone do the job for you — using their skills and their tools? Once you accept this as your plan of attack you'll find that building a repro rod progresses at a rapid rate.

Ah, but being the frugal, quick thinker you are you are already toting up how very expensive this repro rod will be. *Wrong.* How much will it *really* cost you to buy a Model A body at a swap meet, haul it to a body shop for some work, get two repro fenders and two old fenders, rebuild an engine, have a local welding shop or a friend repair a frame, try to hang some late suspension, spend hours trying to find some parts in a wrecking yard and three years later still not have the car anywhere close to running and by that time have lost most (if not all) of your interest in the project? How do you put a dollar value on the hours of frustration we have just described?

If, at this point, you still would like to build a rod, but are unconvinced by the arguments for our plan of attack, perhaps a bit of hot rod history for the uninitiated would be of service.

Early, early hot rods were built without the benefit of any aftermarket help. There was no performance aftermarket hardware. The hot rods consisted of Model T Fords with no fenders, no mufflers and homemade speedster bodies. The aftermarket industry came later with engine parts, two speed rear axles and wire wheels. Although more hardware and more interest was generated with the dawning of the Model A, the real thrust of the "speed equipment" industry was shown immediately after World War II when GIs started getting serious about hopping up flathead Fords. Parts and cars were still plentiful for the hot rodding experimentation without a great investment. For the most part, hardware was limited to the engine, but by 1950 there were aftermarket lowering blocks, shackle sets and adjustable steering arms. Rods were rather unsophisticated, but they served the needs of the rodder. This was truly a do-it-yourself era. Old Fords were everywhere and the wrecking yards contained stacks of engines, transmissions, frames and body panels. If a rodder goofed in Z-ing the frame or in channeling the body over the frame, the loss was small since hardware was plentiful.

As the hardware for old Fords became scarce during the sixties, reproduction hardware came on the scene. This was primarily the domain of the restorer, but the availability of such pieces was not lost on the rodder.

During the sixties, interest in early rods waned due to the advent of Detroit "muscle cars". This waning of interest was a proportionate loss more than made up for in the seventies when renewed interest in early rodding suddenly bloomed. Car clubs, rod runs and early sheetmetal was suddenly "in". Unfortunately, early hardware in decent shape was very difficult to come by and prices soared out of sight. More and more repro hardware appeared — some good, some bad. Slowly, but surely as "gennie" hardware passed from the scene it was replaced by quality "repro" items. The seventies also saw fresh thinking for rods in several areas — suspension, choice of running gear, and increased interest in safety and the building of fully legal, practical cars for daily street operation.

What all of this boils down to is the contemporary rod of the very late seventies and eighties is in actuality a repro rod. It may look like an old car, but it is all new from the ground up — dependable, safe, quick, and responsive, but no less fun or distinctive in appearance than the rod of thirty-five years ago.

Do not take the "plan of attack" lightly. Physically, financially, and emotionally you can figure on living with the construction of your repro rod from one to three years — depending on your enthusiasm, energy and cash available for the project. If you can't accept the above criteria, you ain't gonna believe the second. Namely, what do you have in mind for a street rod? Coupe, sedan, tudor, fordor, fenderless, engine, trans, rear axle, front and rear suspension. The bottom line is you had best buy and read every street rod hardware catalog you can lay your hands on to decide exactly what you need or want in the way of aftermarket hardware to complete whatever project you have in mind. (See the Suppliers List in the rear of this book for catalog sources.)

The absolute worst combination we can think of is not enough time, not enough money and no clear idea of what the finished project will be.

If you don't have or get a full handle on the above by the time you finish reading this book, you'll be lumped into that "95% don't finish" class.

Just as the cars from two eras of rodding are different, so are the methods used to put them together. Your plan of attack in putting together a repro rod should be along the lines of making an assesment on exactly what you want in the way of a car, what is available, what is practical and how much of the car should be done by a professional. That is what this book is about.

THE MIGHTY LIST

On a project of any size — and building a repro rod is a considerable undertaking — it is suggested that a list be made of all that will be needed for the project. This can be most helpful in organizing your plan of attack, estimating costs, ordering parts in relationship to when they will be needed, and so on.

The accompanying list was the starting point for the repro rod depicted under construction in this book. We can assure you it is by no means complete, but may serve as a starting point for your own list of pieces needed.

TCI frame
Buick V6 engine
Turbo 350 trans
Mustang rear axle
Aldan coil-over rear shocks
Aldan front shocks
Walker radiator
Model A grill shell
Mustang steering box
Superbell axle
Superbell spindles
Superbell Mustang brake kit
Superbell brake line kit
Superbell spring perches
Shock mounts-axle
Shock mounts-frame
U-bolt kit for front spring
Spring shackels
King pin sets
Four bar assemblies
Rotors, calipers-Mustang
Steering arms
Tie rod assemblies
Motor mounts
Transmission mount
Transmission crossmember
Mustang master cylinder

Brake pedal assembly
Steel tube brake lines
Wheels
Tires - B.F. Goodrich
'29 Poliform roadster body with rumble seat and Auburn
  dash
Front and rear Poliform fenders
Fender braces
Running boards and molding strips
Running board brackets
Welting
Tilt steering column
Steering wheel
Instruments - Classic
Wiring kit - Ron Francis
Windshield frame
Windshield glass
Wind wings
Wind wing brackets
Windshield posts
Headlights
Headlight bar
Headlight conduit
Rootlieb hood
Hood hinge brackets
Hood shelves
Hood latches
Radiator support rods
Radiator pads
Support rod brackets
Front and rear bumpers, bolts and clamps
Front and rear bumper brackets
Tail light stands with license plate bracket
Tail lights
Body to frame mounting blocks
Door handles and latch assembly
Rumble seat handle and lock
Mirrors
Wood — seat riser, seat back, door panels, kick panels,
  etc. Brad Brown
Gas tank and related hardware
Upholstery
Paint

If you've never built a street rod from the ground up
you should make a list — you can use it to estimate costs,
as a guide to what hardware you'll be needing next, but
more importantly as a "road map" to the kind of car
you're putting together. Hardware changes when the car
is under construction can be very expensive.

*(Right) Fiberglass bodies and new running gear are pretty hard to
hurt, but a fragile item like a radiator should be given some protection when the car is being mocked up. Nothing should be
chromed until the car is taken apart for painting and final
assembly.*

FROM START TO FINISH

The only logical, straightforward way of constructing a
repro rod is to starting with the running gear and work
upward from there — assemble the entire car short of
wiring it. Route all of the plumbing lines, mount the
engine, transmission and rear axle, install wheels and
tires, the radiator, all steering and suspension
components, do all fabricating and welding, make sure
the body is bolted to the frame and the doors and other
panels fit the way you want them to. When you are
completely satisfied the car looks like you want it to in
terms of rake, wheel/tire combination, etc. take the car
completely apart and paint or plate each individual piece
and then re-assemble the car. This is the proven method
of rod building. Attempts to "short cut" this plan will be
met by parts that don't fit, chipped paint and frustration
and disgust with the entire project.

If you are going to farm out major segments of the
construction such as wood, upholstery and paint, get
these people in on the project long before they are
needed and encourage them to communicate with each
other. For instance, let's say you plan to have a local
cabinet shop do the woodwork in the car. He should meet
the upholsterer at the car before going to work. Perhaps
the upholsterer would rather build his own door panels
and kick panels — for any one of several reasons. Why pay
to have the job done twice or force the upholsterer into
working with panels too thick or too thin for the particular
way he upholsters? How does the seat back get attached
to the body? The upholsterer may have some very strong
thoughts about this — logical, sound reasons for doing it
in a particular way. Maybe the upholsterer wants to do the
wood construction also. Ask.

Maybe you plan to prep all of the body panels yourself
to save money, than take them to a painter. In some cases
this can save hundreds of dollars, but find out *exactly* how

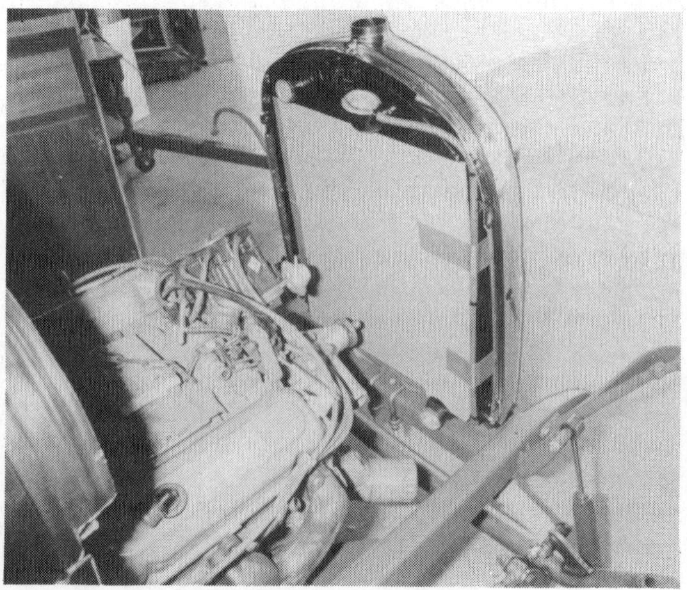

the painter wants the body prepped prior to having him shoot the color. Then prep one small panel — a door, a dash or a rear fender — and take it to the painter to see if he's happy with your work. Building a repro rod can be a nightmare if you don't plan ahead. Even as you are buying the frame to start the project you should have some idea about how you plan to register the vehicle. Planning of this sort avoids endless frustration.

(Right) Although you need not bother with final hood and door alignment when the car is first assembled, the windshield frame should be set in place to insure the support posts are correctly aligned on the body. Tires and wheels should also be installed. Now is the time to make changes.

After all fabrication and welding is completed, take everything apart and paint all of the pieces. Then reassemble. This chassis is now ready for the body.

With the body painted in pieces and rubbed out, the body can then be assembled on the running gear. An extra pair of hands is mandatory here to prevent chips and scratches.

(Left) If you can hold your pants up without a belt — do so when working on the car after final assembly. You'd be surprised how quickly a belt buckle can dig into lacquer. Scraps of rebond foam rubber carpet padding is another way of helping to prevent chips and scratches caused by dropped tools and various nuts and bolts.

Frames

Model A, '32 Ford and '33/'34 Ford reproduction frames are now available in various stages of completion from various sources. As this is written one manufacturer builds all of the repro Model A frames and sells them in various stages of completion to a variety of sources who in turn add their hardware or simple resell the hardware they have purchased from the original source. The same is basically true for the other two types of frame.

Our recommendation on buying a frame is to purchase the frame as complete as possible from one source. "Bargain hardware" from several sources normally has a way of turning into an expensive nightmare. For instance, let's say you buy a four bar assembly for a Model A frame from one source and mount the brackets to the frame. Oddly, a steering box mount from another source interferes with the four bar bracket on the left side of the frame. Upon checking around you discover that a steering box bracket made by someone else will not interfere with the previously installed four bar. This bracket is then installed by welding it solidly to the frame. Unfortunately, it is discovered some time later this last steering box bracket which solved one problem, now creates another one in the form of interference with another brand of motor mount.

Situations like the one just described are not unusual

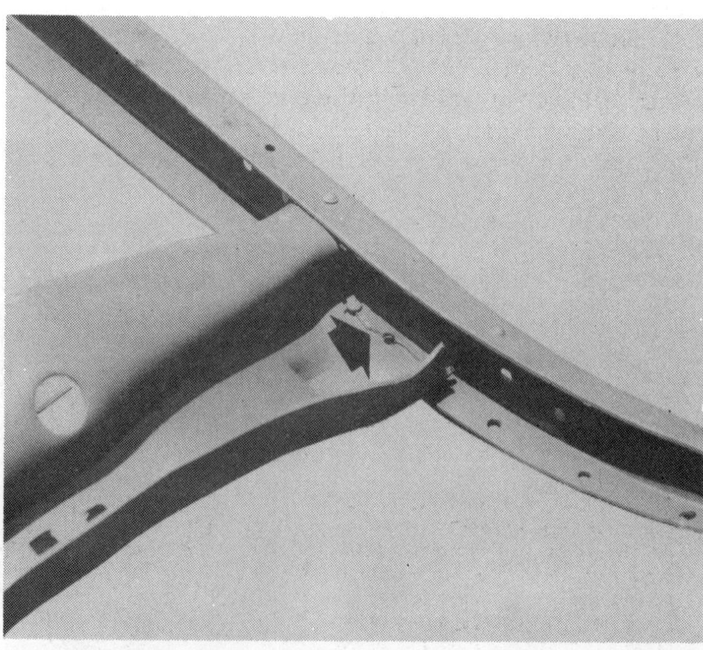

Original Model A frames are now fifty years old and a lot of them were driven long distances on bad roads. Cracks and bends can be repaired, but even when correctly executed an old repaired frame will never be as good as a new boxed, reproduction item.

when an amateur rod builder gets in over his head and tries to save money at the same time. Variations of the above type situations literally can go on forever as mistake after mistake is made in order to try to correct one error that should not have been made in the first place.

Study the hardware catalogs and make a determination of exactly what you want in the way of running gear from one end of the car to the other — before ever buying the first piece of hardware. If you are a fabricator or machinist you might be able to handle problems without creating new ones. You should know what engine, transmission, rear axle and brake assemblies you will be using before buying any hardware. It is wise to acquire this running gear very early in the project. Let's create another little scene just to show how a change in plans can create problems. Let's say you order a Model A frame set up for a small block V8 Chevy and Turbo 350 transmission. The motor mounts and transmission crossmember are welded in place. You have painted the frame and all related hardware and mounts. Without warning you get a chance to buy a small block Ford V8 and C4 trans — both complete and in very good condition for far less money than it is going to cost you to acquire comparable GM components. If the Ford running gear is acquired, the motor mounts and the crossmember will have to be removed and replaced. If you are not equipped to do the job, the frame must be transported to someone who is able to do the job. Little delays like this have a way of turning into big delays. A setback that first appears to delay the project only a few days suddenly turns into a couple of weeks, then months, and as change after change is made, the years roll by and the discouragement grows.

Naturally the argument against having one supplier do a complete frame with all brackets and sticking to a prescribed plan is that initially the cost is prohibitive. This can be doubly hard to justify if you are a fair fabricator and think you know more than most about what you want in a

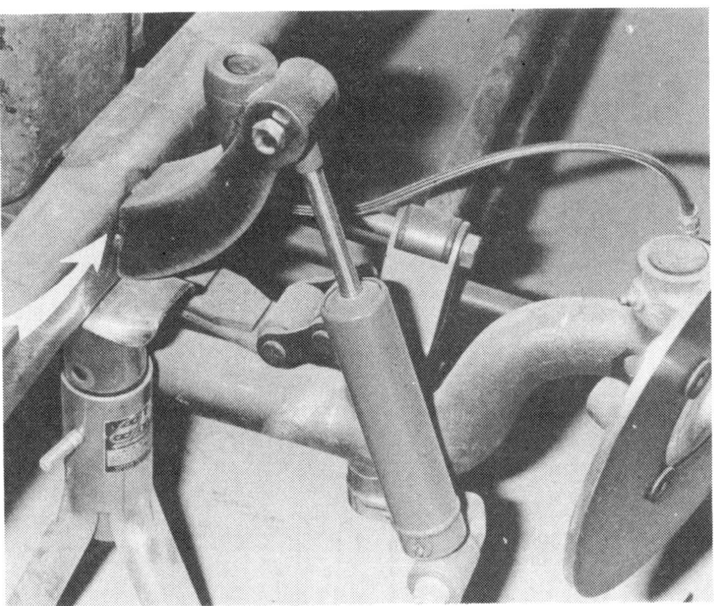

If you start with a bare frame and add up your own brackets or buy brackets from another source, the entire running gear of the car should be assembled with all brackets tack welded in place. This is extremely important. For a first time builder the little problems that crop up can be a nightmare. Keep in mind that if you are building a fenderless car that a lot of this hardware will be exposed and that your design and execution will be seen by a lot of rodders.

hot rod. What then? With a slight modification, I would recommend the same basic approach. Buy motor mounts, crossmember, master cylinder mount, four bar, shock mounts, etc. from only one manufacturer. Tack weld everything in place, completely assemble all front and rear suspension.

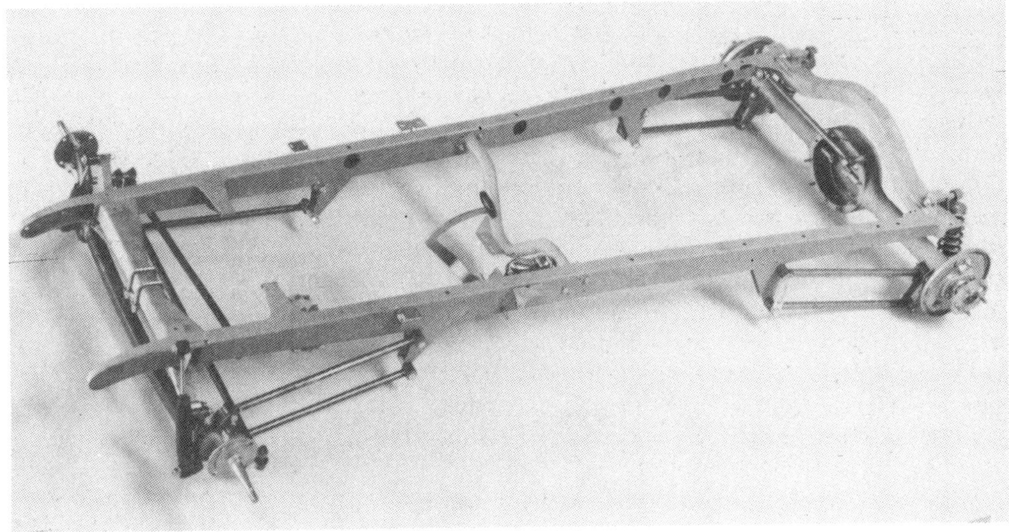

TCI Engineering manufactures this reproduction Model A chassis which can be purchased this complete or as a bare bones frame. In addition to the hardware they manufacture, TCI also specifies what other components they are using such as Super Bell axles, spindles and brake hardware, Deuce Factory stainless hardware, Mustang or Vega steering and so on.

Even if you buy a complete chassis from one supplier chances are that you'll still have work to do to get the fit and finish you'll want for your rod. These small tabs have been added to further anchor the front splash apron.

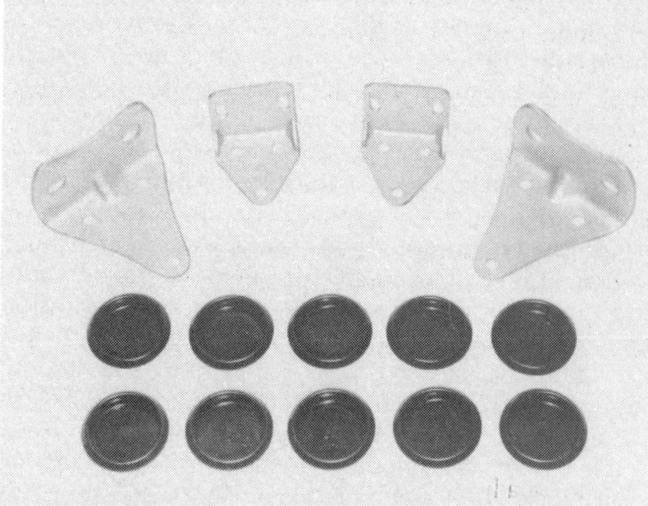

With their complete chassis, TCI even includes little items like the stamped body mounts, hood latch brackets and even the nylon plugs for the body mount access holes.

Bolt on the tires and wheels and set the body in place, but don't bother to align the doors, the hood, etc. Are the wheels and tires centered in the wheel wells? Does the car have the rake you want? Does the master cylinder clear the floorboard? Does the brake pedal interfere with the steering column? Does the exhaust system interfere with the body, or rear bumpers? The list of questions goes on. The point is that with all running gear either bolted or tack-welded in place, the problems can be readily solved. Once all of the fitment problems have been solved, complete all welding on the running gear.

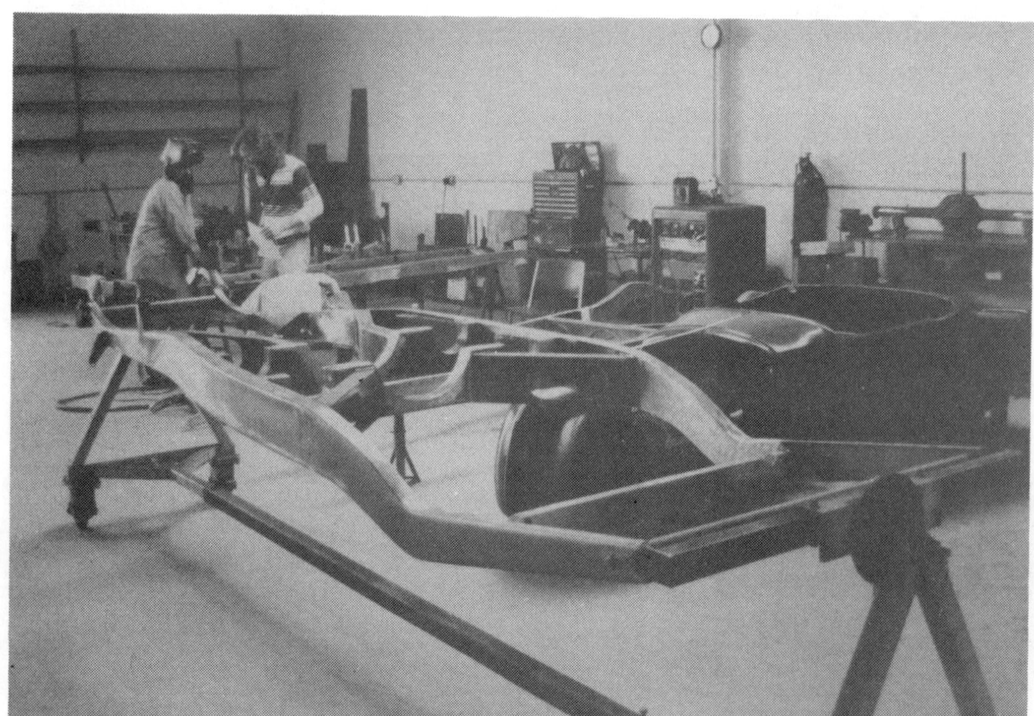

A professional chassis shop has all of the equipment — plus the experience — to do the job right the first time.

The reason you shouldn't spend time aligning the hood and doors with the chassis tack-welded together is that the frame can move around during the final welding process and fine adjustments will have changed slightly. Even a frame welded in a jig can move slightly when removed from the jig.

Probably the vast majority of failures in rod building (incomplete cars) lies with the fact that the builder is constantly changing his mind about what he wants with first one piece of running gear and then another.

By now the message should be clear. Form a simple, straightforward, workable plan for you and stick to it.

THE HIBOY

Traditionally, a "hiboy" is a term in street rodding referring to a Model A Ford roadster body being used on a '32 Ford frame or a '32 roadster on a '32 frame — in all cases the car must be without fenders or running boards so the graceful lines of the '32 frame are fully exposed. Naturally, building a hiboy with a '32 body on a '32 frame presents no particular problem; but Model A roadsters on a '32 frame is another kettle of fish and you should know this before plunging into such a project. To properly pull this off, the '32 frame rails must be pulled inward to meet the curvature of the bottom of the Model A body —which obviously is not the same as that of the '32 body. Heating the frame rails and pulling them inward with a variety of tensioning devices is not particularly difficult, but keeping the frame square and the frame rails from twisting in the wrong direction or in the wrong place — or both — is beyond the skills of most hobbiest rod builders.

The recommended route to take when building a hiboy using a Model A body is to buy a repro '32 frame which has been modified by the manufacturer to accept a Model A body. As this is being written, TCI is such a source. Presumably there are others.

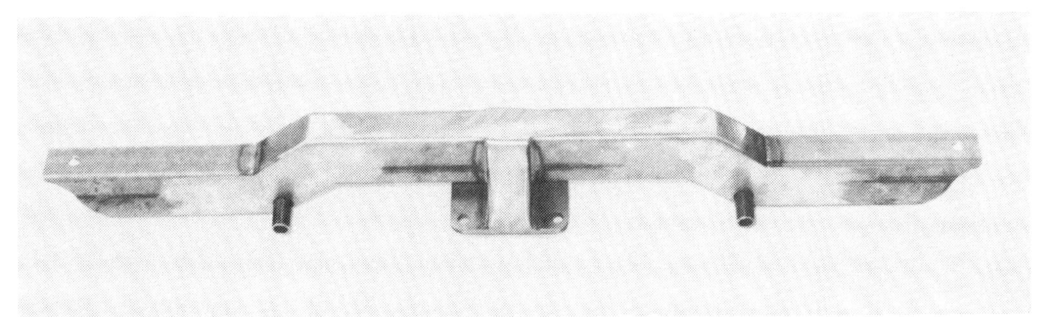

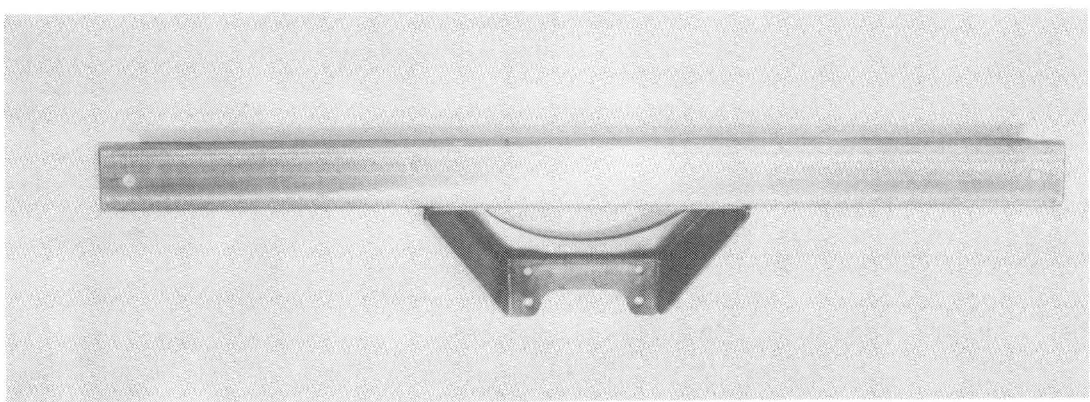

(Top and Above) Most chassis shops can provide the correct cross-member for Corvette or Jaguar rear axles. Make sure you fully understand what else may be involved to complete the installation such as shock brackets, link bars, brackets and so on.

Front Suspension

More history is perhaps in order to explain how we've arrived at contemporary rod suspension. If fading memory will serve me once again, all Ford passenger car suspension from the era of the "T" through 1948 had a wishbone type suspension. A wishbone was a large V-shaped piece of steel. The apex of the V was a ball joint arrangement that attached to the running gear of the car in the area of the bellhousing. The "arms" of the V ran forward and attached to the solid front axle in the area of the attachment points of the transverse leaf spring pack to the axle. Not a bad arrangement for the time. Simple, easy to produce, lots of wheel travel for the bad roads of the time, durable, and easy to service. No doubt about it, wishbones had a lot going for them in reasonably stock vehicles.

Wishbone type front suspensions normally created problems only when hot rodders entered the picture. In simplified terms, the rodders jerked out Henry's four bangers and installed flathead V8 (later overhead valve V8) engines and their accompanying transmissions. This running gear had no provision for mounting the apex of the wishbone. This shortly became little bother for the rodder, who split the wishbone, did away with the central pivot point and attached the two remaining ends to the underside of the frame.

Splitting the wishbone solved the problem of getting

the later running gear into the early chassis and created the problems now associated with all split wishbone type suspensions. When an axle held by a split wishbone moves up and down, caster changes. With a stock wishbone the entire axle assembly pivots from one point — the single ball joint at the apex. Caster change causes no problem, but with a split wishbone, the axle is now pivoting from two separate points. This means if one

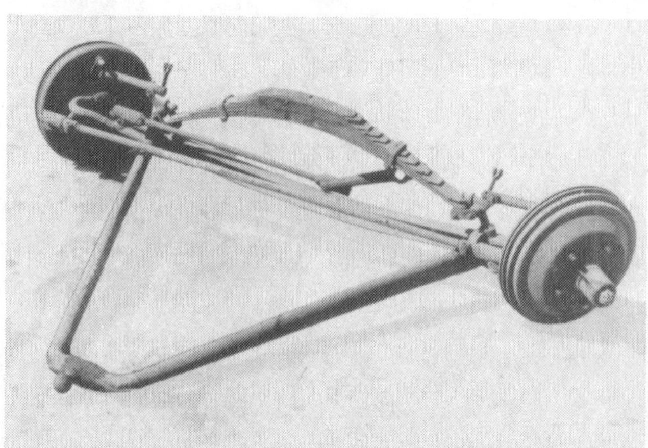

This is a stock early Ford wishbone, leaf spring, tie rod, drag link and axle assembly mentioned in the text. All of this pivoted from the ball at the apex of the wishbone.

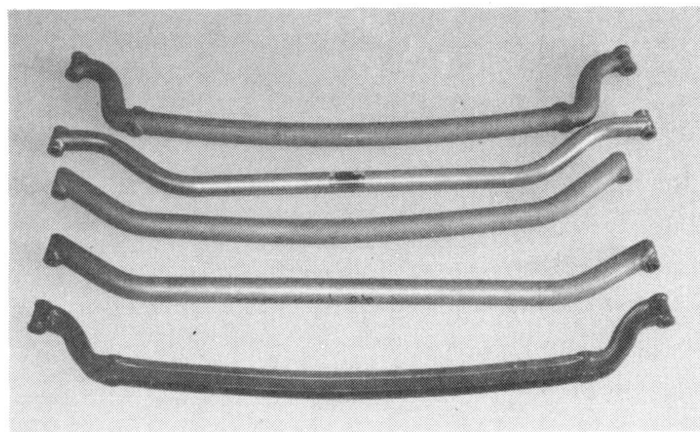

This is the best evidence we could come up with as to why you should always shop quality and not price. Some of these axles could create problems that no one could solve.

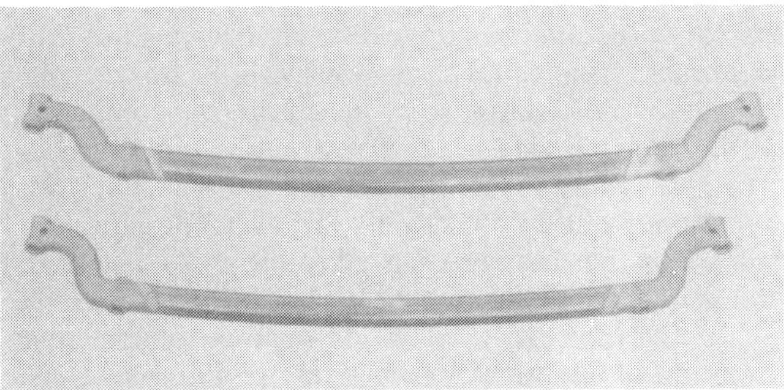

These two axles by Super Bell are the most popular in the repro rod building industry today. Of these, the favorite is the 4-inch drop unit at the top. The lower axle features a 3-inch drop.

wheel moves up or down the caster will change for that wheel. Because the two wheels are connected by a beam axle, the wheel moving up and down will affect the other wheel by attempting to bend the axle. This creates some weird ride and handling sensations even if the car is equipped with a forged I-beam axle, which will bend under loading.

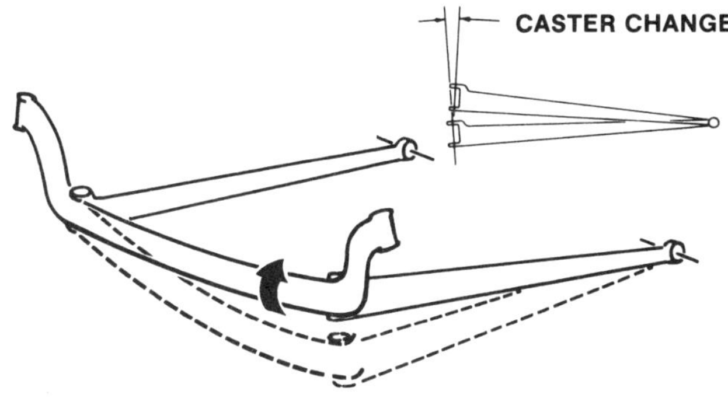

Drawings courtesy of Pete & Jake's

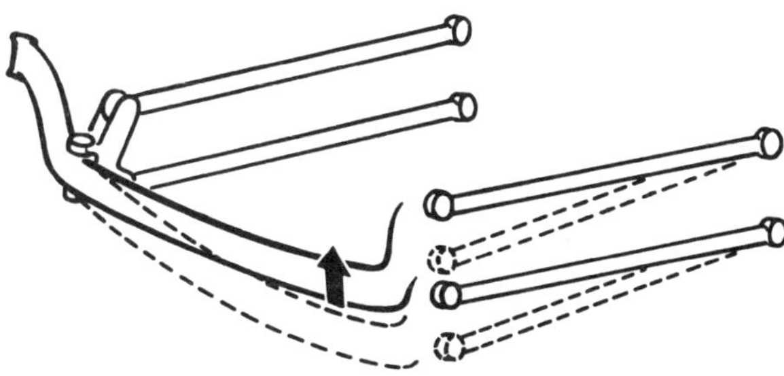

This situation was simple tolerated and no one did much to correct the situation until well into the 1970's. due to the scarcity of forged I-beam, stock type front axles to put under drag race and street driven cars, fabricated, round tube axles came into being. A round structure is much more resistant to bending than an I-beam shaped structure. Almost overnight the rodding world went from a forgiving, geometrically incorrect front suspension to an unforgiving, geometrically incorrect front suspension. Breakage began to occur routinely as more and more round tubular axles came into use.

Rodders have always been quick to use the "bigger hammer" approach to engineering and began fabricating massive brackets to retain the ends of the split wishbones in an attempt to eliminate the breakage. They succeeded in many cases. In time their success led to frames and crossmembers cracking. If that can be considered "good news," the "bad news" is the ride and handling of a split wishbone car was never quite right. There were strange noises as loads went fromt split wishbones to axle to frame and then into the body. The answer to the solution came in a design which has become known as 4-bar suspension. To be legally correct, 4-bar is a registered "trademark" name for the parallel link radius rod kits manufactured and sold by Pete and Jake's. To be practically correct, there are a number of manufacturers making four bar, parallel link radius rod kit.

Four bar design front suspension utilizes the principle of the parallelogram — when one end of the figure moves in relation to the other, all sides remain parallel. Connecting a parallelogram to each side of the axle and each side of the frame allows the axle to move up and down without caster change. Thus there is no attempted twisting of the axle, no unengineered loads being fed into the frame, no weird handling characteristic or noises being transmitted into the car.

Simply stated, the four bar type of suspension is the current favorite for a contemporary street rod. The geometry is simple, straightforward and workable; the hardware is readily available at reasonable cost. All other suspension pales by comparison for the hobbiest.

Typically, a four bar suspension kit will consist of axle

brackets, frame brackets and the four adjustable bars with bolts, lock nuts and bushings. Some kits have the left side frame bracket fabricated to serve as a steering box mount. In an attempt to clear up some confusion we'll point out that the four bar axle brackets are often referred to as "batwings." These pieces of hardware are available from many manufacturers in a wide variety of designs and quality. Batwings range from well thought-out stamped, formed, heliarced mild steel or beautifully cast to poorly designed, terribly executed flame-cut pieces of trash that no amount of rework can save. Among other things you must keep in mind when buying front suspension components is that they all have to work together. Thus when buying a four bar kit, you should specifically ask if the hardware will be compatible with all of the rest of the front suspension hardware. For instance, some batwings will interfere with the tie rod because the crotch of the 'wing is not deep enough. This greatly reduces the turning radius. To a great extent the spindle dictates the type of brake hardware you'll be using and the steering arm that can be fitted to the spindle.

As you begain to formulate what you want in the way of front suspension, you will run across reference being made of batwings being available for 2-inch axles and for 2¼-inch axles. These dimensions refer to the width of the spring perch boss on the axles — thus if you have 2¼-inch batwings, the axle must be of the same dimension.

Drawing courtesy of Pete & Jake's

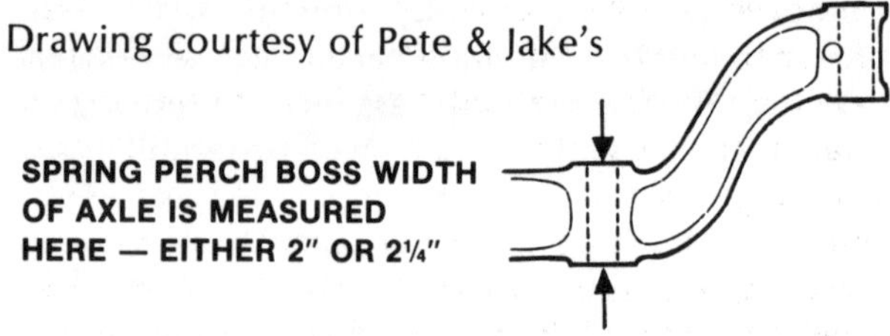

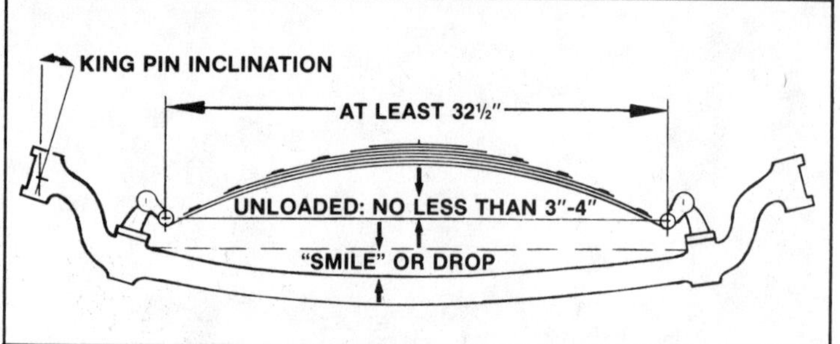

It is important to understand that in most cases the spindle arms will have to be heated and bent up or down (depending on year and make of spindle and amount of drop of axle) to center the tie rod in the crotch of the batwing for maximum turning radius. Superbell reproduction spindles require their reproduction bolt-on arms which they have prebent for certain applications.

AXLES

Super Bell axles are available in 3 and 4-inch drops. If you live in a rural area or plan on driving on rough roads or you simple don't want much front rake to the car, then the 3-inch drop is probably the best selection. The 4-inch drop is by far the most popular. It is practical from every standpoint for the great majority of rod builders, it creates no unusual problems, and when used in conjunction with a small tire gives the front rake known as the "California look."

When assembling all of the front suspension components you can expect the spring shackles to be roughly parallel to the ground when the front suspension is simply hanging.

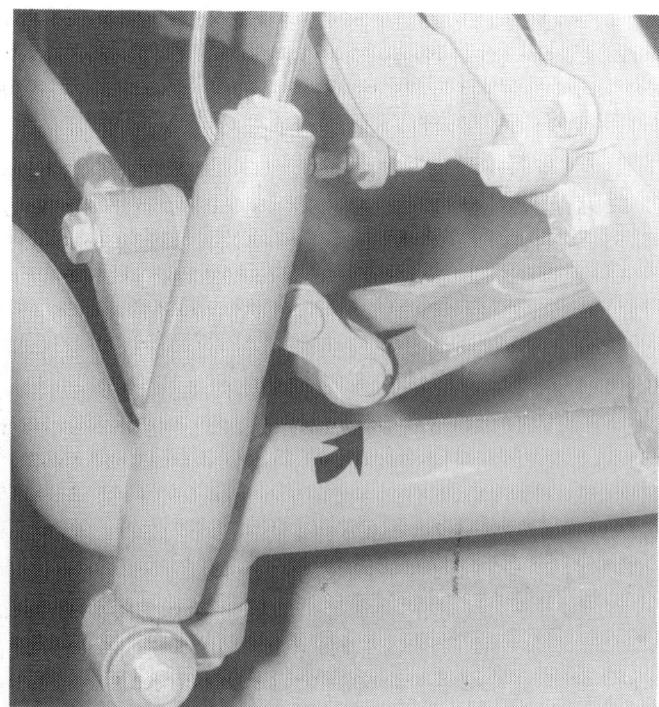
When the front suspension is properly set up and the front suspension is in a static, loaded condition, the shackle should be very close to a 45-degree angle to the ground.

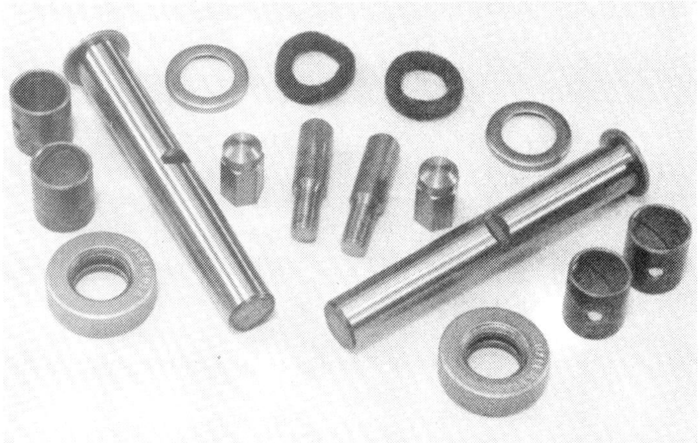

If you are looking for durability, replacement king pin sets with grooved brass bushings are hard to beat. This may not be fancy or space-age, but it sure does work.

Super Bell manufactures these spindles from forged 4130 steel. Dimensionally, these spindles can be used to replace early Ford spindles, but they should not be referred to as reproduction spindles because the steering arm assembly is a bolt-on. The firm makes a variety of steering arms specifically designed for use with the spindle.

From a suspension engineering standpoint, it should also be kept in mind that the more an axle is dropped, the more it will want to rotate. This is because the moment arm is greater.

Finally it should be pointed out that although there are thousands of street rods and light trucks running around with a dropped I-beam front axle, the fact remains that they are inferior to a quality tube axle. The dropped I-beam will flex and bend and it is only when the axle stops bending that the steering hardware can get on with the job of steering the car.

Installing a four bar front suspension for the first time can test the patience of many a rodder. To the novice it seems that all factors must be taken into consideration at once and nothing can be accomplished until something else is taken care of — hopefully, these installation tips from Pete & Jake's will help ease the mental anguish.

When an axle moves up and down it also moves back and forth — or front to rear on an arc related to the pivot of the wishbone or bars holding it. Obviously, the shorter the bars the tighter the radius the axle travels, and thus more movement. This front-to-rear movement is very slight when the bars are parallel or near parallel to the frame because travel in the arc is at a minimum. In the 2-inch up, 2-inch down travel required by a contemporary rod, the axle will move approximately 1/16-inch front to rear. This causes no problem. The problems occur, and are blown out of proportion in the mind of the rod builder, when assembling and disassembling the front end. When the axle is dropped down from ride height by four inches, the axle will move back ⅜ to ½-inch.

When a stock wishbone set-up is used the axle also tilts or rotates forward (negative caster) when the axle assembly is dropped down and the uncompressed spring slides easily in and out of the crossmember. When a parallel bar system is used the axle does not rotate forward (caster remains constant) as it drops down away from the frame. This causes the spring to bind.

When setting up the front end in a frame with no engine or body weight and using a full spring (few or no leaves removed) in the crossmember, the axle will be dropped down from ride height. If the bars are connected and adjusted under these conditions, they will actually be longer than necessary when the car is fully assembled and set at ride height. Stated another way, if the front end is set-up with a full spring in the crossmember with no weight to compress it, the axle will move forward when the full weight of the car is lowered down on the spring, causing it to bind.

The proper way of setting up the front end is to position the axle assembly and adjust the bars at ride height. To do

Regardless of whether Super Bell spindles or old Ford spindles are used they must be fitted with king pin bushings, reamed and then fitted with king pins and related hardware.

this use only the main leaf (and possibly the second leaf) mounted on the shackles and slipped up into the crossmember. This simulates the full spring for positioning the axle but allows the bare frame to rest at ride height. A good rule of thumb is have 3 to 4 inches between the center of the crossmember and the axle for ride height. With only the main and the second leaf installed in the crossmember, slip a short length of 2 × 4 lumber on its side between crossmember and axle to support the frame. At this point the bars can be adjusted for length, squaring-up the axle and setting caster angle. Leave the front end assembled this way until final assembly when the full weight of the car is assembled on the frame. The final step will be to disconnect the bars from the frame brackets and install the full spring. Then lower the car to the ground on the spring compressing it to ride height. At this point the bars can be reconnected to the frame brackets and the assembly is set.

If the full spring is mounted on the front end and the bars are hooked to the frame brackets, the spring will not go into the corssmember without brute force. The spring must be compressed before it and the bars will line up. You might need only to disconnect the two upper bars, letting the axle assembly rotate forward in order to slip the spring into the crossmember. Ride height will vary from one vehicle to another depending on the spring (reversed eye, stock eye, number of leaves and arch) in addition to the total weight riding on the spring, so fine tuning the length of the bars is mandatory for each car. If the front end is installed with the spring in a bind, it will not work smoothly and freely and will put a strain on the shackles.

SPRINGS

Don't buy a spring until you have the axle in hand that will be used. As the instructions from Posies indicates, measuring your axle for a spring is relatively simple.

Measuring For A Spring

Step One

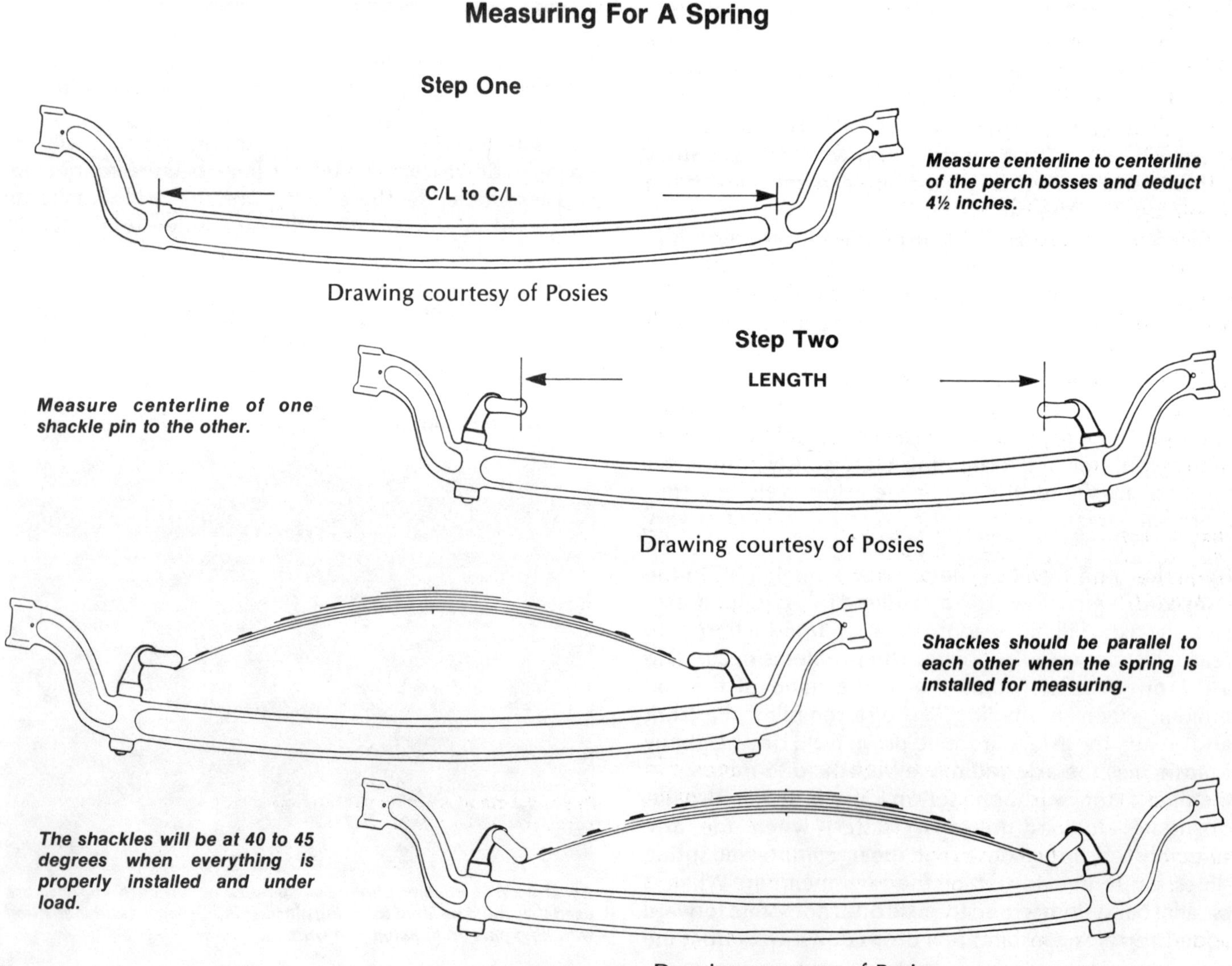

Drawing courtesy of Posies

Drawing courtesy of Posies

Drawing courtesy of Posies

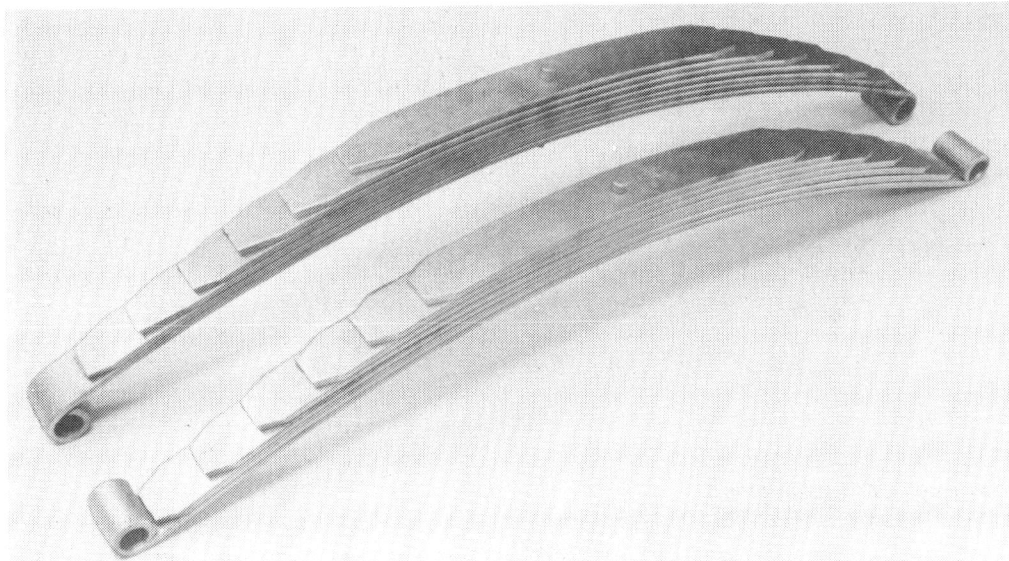

Here's the difference between a stock spring package and a reversed eye package. The stocker is at the top. The reversed eye assembly at the bottom will lower a car approximately 1 inch if everything else remains the same.

Measure from centerline to centerline of the spring perch bosses and deduct 4½-inches. This gives you the size of your spring. If you have axle, spring perches and shackles in hand then assemble the hardware and position the shackles so they are parallel to each other. Measure from the centerline of one shackle pin to another. That's the length the spring should be.

The shackles should be parallel to each other when the spring package is installed but under no load.

The shackles should be 40 to 45 degrees when the spring is installed and loaded.

Over the years rodders have added a lot of refinements to the old Ford style transverse front spring. In the long run, probably the most important modification to be made is the rounding and smoothing of the ends of each leaf with a grinder to keep the end of the leaf from digging into the one below. The second biggest improvement to be made to a front spring to improve the ride is to install strips of Teflon between the leaves to greatly reduce the friction between the leaves and allow a much softer ride.

Round lower edges of spring leaves so that they will slide easily as spring flexes. Polypropylene between leaves helps sliding action of leaves.

Drawing courtesy of Pete & Jake's

Posies has a spring package having dimples stamped in the ends of the four longest leaves. These dimples contain plastic which is captured between the leaves. Like the strips of teflon, this also provides a superior ride.

Alter the length of the leaves within the spring package as shown for optimum ride. Steel spacers will have to be added to the top of the spring package to make the spring

thick enough to be clamped tightly in the frame.

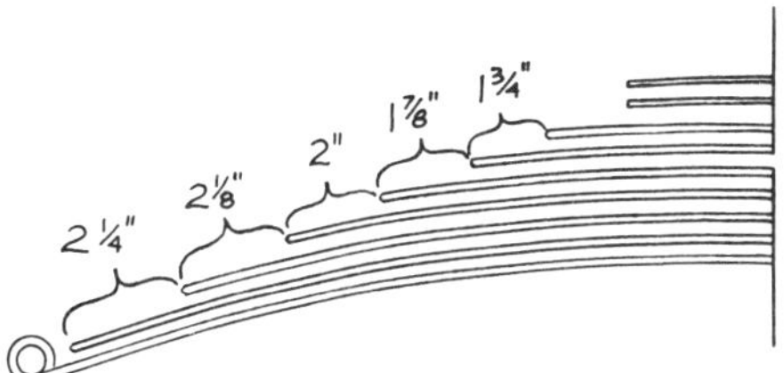

Distances between ends of leaves are shown in drawing for best ride. Add spacers on top of spring as needed to fill out for fit in crossmember.

Drawing courtesy of Pete & Jake's

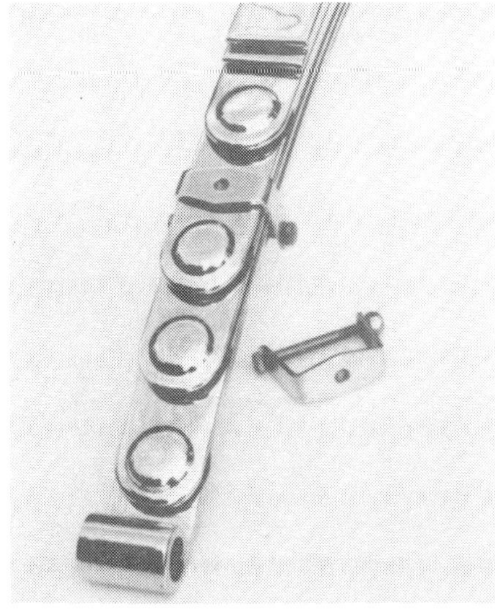

This is the dimpled spring package offered by Posies which contains plastic inside the dimple.

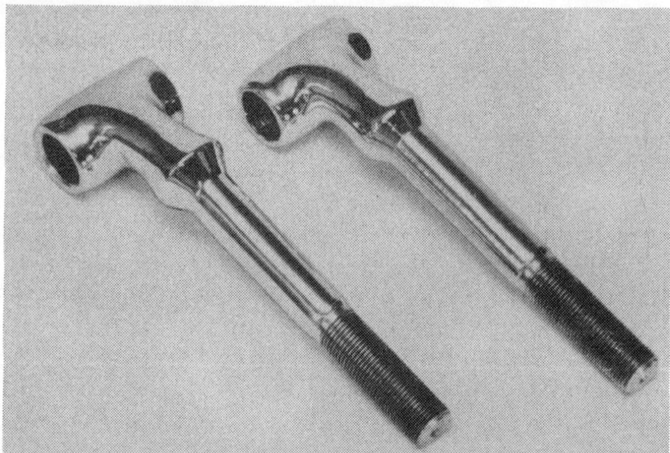

Spring perches for early Ford hot rod type suspensions are available in standard or extra long shank. The latter version makes a lot of sense because they can be teamed up with . . .

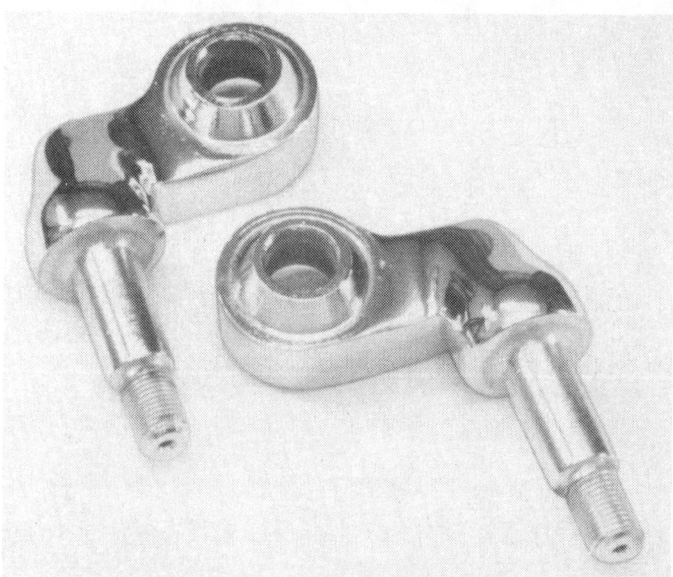

. . . the Super Bell bolt-on lower shock mount which is designed to become a part of the axle, bat wing, and spring perch assembly . . .

. . . for a front suspension that appears clean and functional — and is.

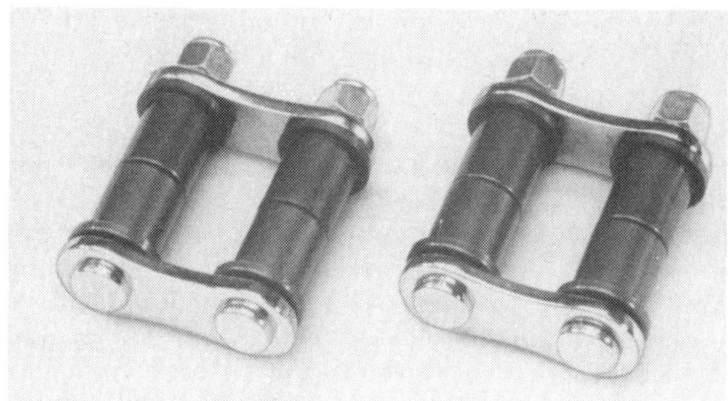

Shackles are available in several widths and several bushing diameters. Pete & Jake's has ten different part numbers to cover all of this.

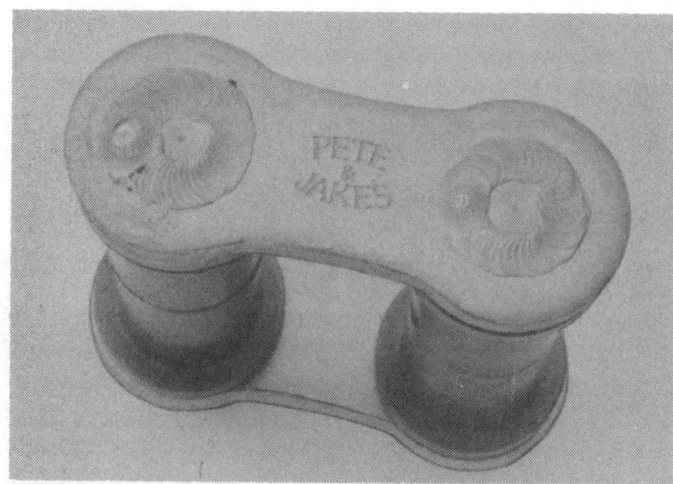

Flat faced shackles such as this item are usually required in hot rod front suspensions in order to clear the shock body.

The Deuce Factory makes a full line of stainless steel front suspension components including four bar adjusters available in two different angles and straight in addition to the beautifully polished stainless batwing.

STEERING

Like all other component systems on a hobbiest-built vehicle, look long enough at steering systems and you'll eventually see anything you can imagine — and a lot of things you can't imagine! Like brakes, steering is a very unfunny areas of hardware.

If you'd rather be safe than sorry and want to be as simple and straightforward as possible in putting a steering system under a repro rod, there are really only two systems to consider at this point in the history of rodding. The two systems are generally described as Mustang steering and Vega cross steering. We are indebted to Pete and Jake's for the information from their catalog which amply explains what you need to know about both types of steering.

MUSTANG/COUGAR STEERING

The design of the '65-'73 Mustang/Cougar steering box makes it a natural for early Ford installations as it offers convenient mounting and a sector shaft long enough to go beneath a boxed frame. The boxes are readily available in wrecking yards and from dealerships. There are two ratios offered in the FoMoCo steering box — 4 turns lock to lock (most common) and 5 turns lock to lock. The latter box is the slower steering of the two. A removable column on the '69-'73 boxes also allows the adaptation of tilt

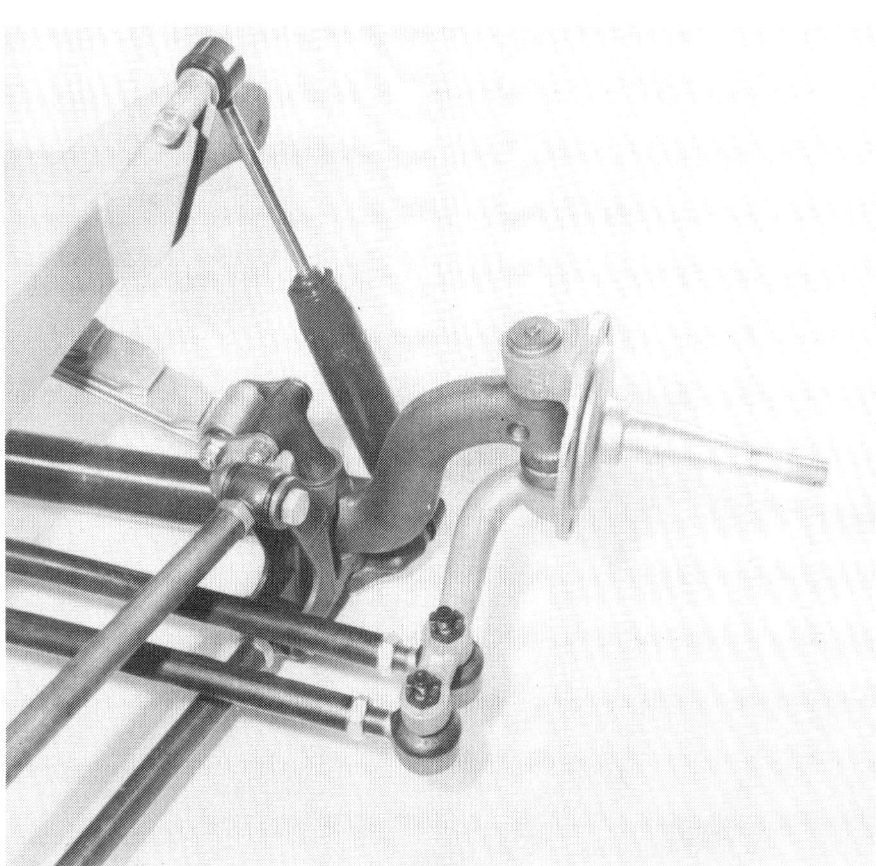

Cross steering is very popular on hiboys because the pitman arm and draglink are hidden under the chassis. This requires a tie rod arm (steering arm) on the right side with two eyelets — one for the tie rod and one for the drag link.

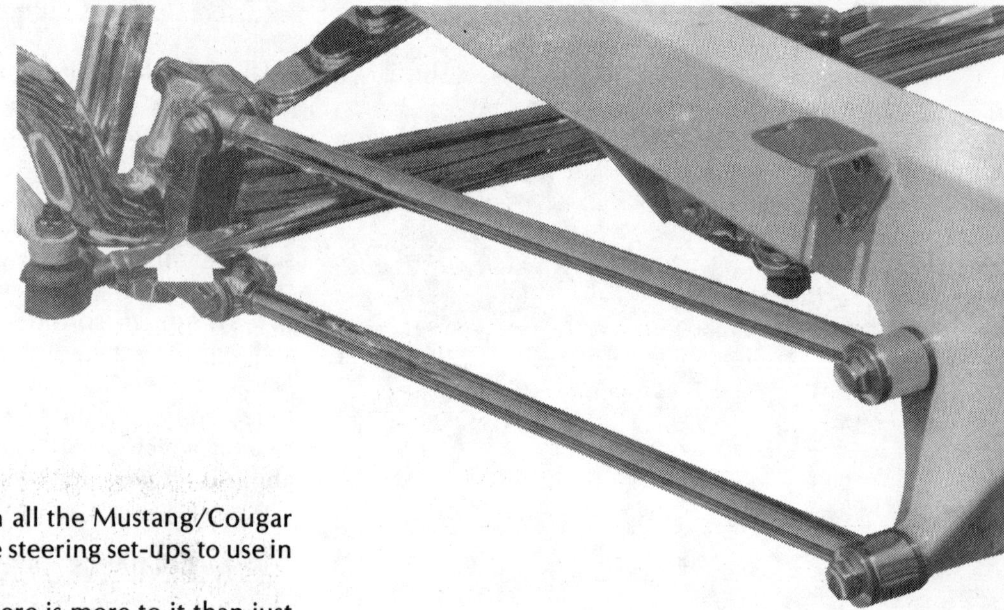

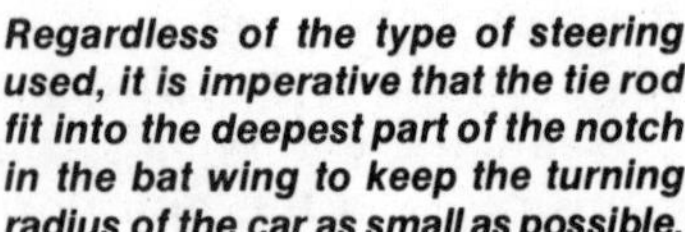

Regardless of the type of steering used, it is imperative that the tie rod fit into the deepest part of the notch in the bat wing to keep the turning radius of the car as small as possible.

columns from other cars. All in all the Mustang/Cougar box is one of the most favorable steering set-ups to use in a repro rod.

In any steering installation there is more to it than just mounting the box and connecting the drag link. What appears to look good and work right when the chassis is sitting on stands in the garage may in fact be all wrong. Arbitrary placement of steering components can mean incorrect steering geometry and serious handling problems once the car is on the road.

Other than being hard to steer, hot rods build twenty or so years ago didn't seem to have major steering problems. The reason for this was the front end design and steering set-up was left basically stock. Even if the car was equipped with a dropped axle and the very popular '49-'56 Ford pickup steering box, the steering geometry had not changed enough to cause trouble. Then rodders began installing the Mustang steering gear, and what seemed to be a great improvement only made the car handle dangerously strange. Any deviation in road surface would turn the car in one direction or another. This action would also turn the steering wheel and thus give the driver the feeling he was not the only one steering the car at all times. This behavior was promptly labeled "bumpsteer."

Bumpsteer results from incorrect steering geometry which causes the car's steering to change direction during vertical suspension travel. In the design of the Mustang/ Cougar box, the pitman arm is opposite that of the early Ford steering box in that it rotates above the sector shaft instead of below the sector shaft. This drastically changes the drag link angle and steering geometry in relation to a stock style front end using either stock or split wishbones or a single pivot type radius rod.

Bumpsteer can be eliminated from a Mustang/Cougar steering installation if a parallel radius rod (four bar) setup is used. The parallelogram design features of the four bar system keep vertical suspension movement of the axle and spindles relative to the movement of the drag link. Keep the drag link parallel to the four bar system and bumpsteer is eliminated.

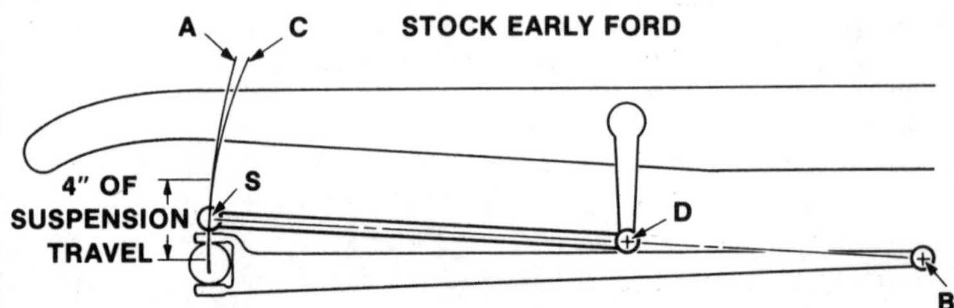

Drawing courtesy of Pete & Jake's

There are three basic points that determine the steering geometry of an early Ford front suspension. They are the pivot points at each end of the drag link, and the point from which the front end assembly pivots. Point S (see drawing) is the center of the ball inside the rod end attached to the spindle steering arm and the drag link. During vertical suspension movement, point S, as part of the axle/spindle assembly, travels in an arc (A) centered at the wishbone pivot point (B). Point S also travels an arc (C) centered at the other end of the drag link (point D) which is connected to the pitman arm. Although point S must travel 2 different arcs at the same time, these arcs will be very close for a limited amount of vertical movement if points S, D and B are all on the same centerline. In other words, steering geometry will be optimum if the drag link (S-D) is parallel to the imaginary link (S-B).

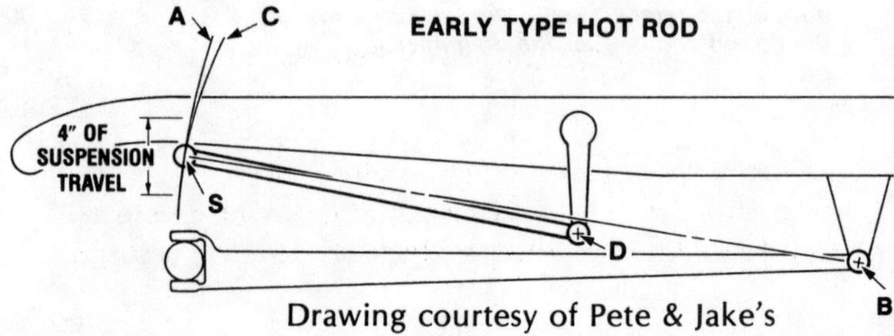

Drawing courtesy of Pete & Jake's

Modifications to the stock suspension cause the steering connection points and the geometry to change. In this case, point S has moved up four-inches vertically with the installation of a dropped axle and reversed eye spring. Point D has moved up only a little because a shorter pitman arm is used with the pickup steering. However, point D has also moved 1 to 1½-inches away from the centerline S-B. Because the drag link (S-D) is no longer parallel to the imaginary link (S-B), arc C will not follow arc A as closely and point S will be forced to move back and forth as it travels both arcs. Fortunately the difference in the arcs is still very slight within the limited amount of vertical suspension movement and bumpsteer, if any, will be insignificant.

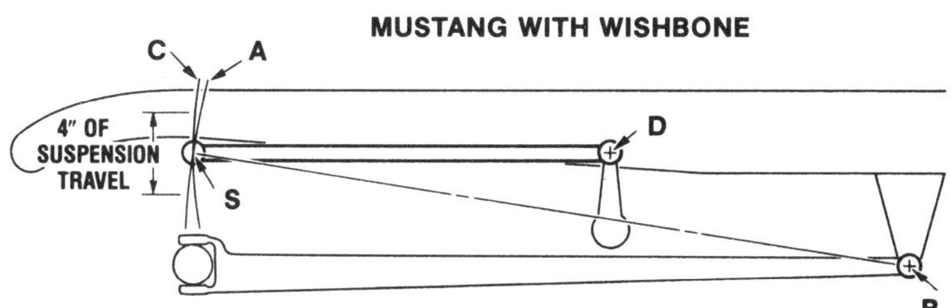

Drawing courtesy of Pete & Jake's

The installation of Mustang steering using a wishbone (single pivot) set-up causes a drastic change in steering geometry. The unsuspecting builder has been told that correct steering geometry can be attained by mounting the drag link parallel to the wishbone. This is not true. As shown in the illustration, the drag link (S-D) should be parallel to the imaginary link (or centerline) S-B. Due to the steering box design and the position of the pitman arm, point D has moved a considerable distance from the centerline S-B which means that arcs A and C will no longer be close even within the limited amount of vertical suspension movement. This in turn forces point S, the steering ball on the spindle arm, to move a great amount as it travels both arcs. This results in bumpsteer.

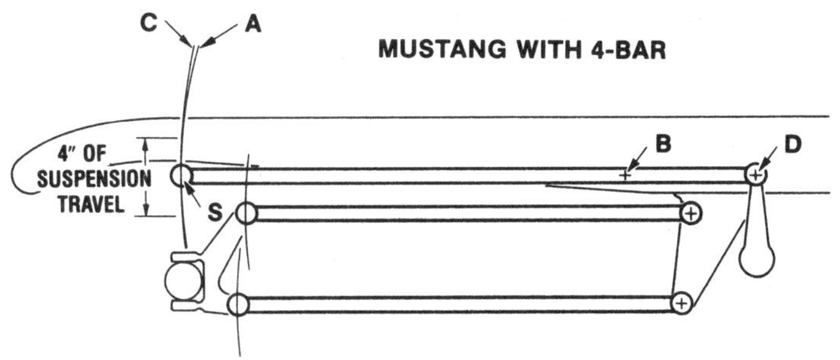

Drawing courtesy of Pete & Jake's

Mustang/Cougar steering installed with a four bar system offers an ideal steering geometry which eliminates bumpsteer. Based on the parallelogram, point S is part of

the axle/spindle assembly and travels an arc (A) equal to the length of the parallel radius rods, centered at point (B) parallel to the rods. Point S also travels an arc (C) centered at the other end of the drag link (point D) which is connected to the pitman arm. If the drag link (S-D) is parallel to the imaginary link S-B, arcs A and C will be close enough within the limited vertical suspension movement to not cause bumpsteer. If the drag link is not parallel to the radius rods, points S, B and D will not be on the same centerline as in illustrations #2 and #3, and the resulting bumpsteer will again depend on the degree of angle difference.

(Above and Below) Over the years anything and everything has been tried on hot rods in the way of front suspension hardware. Here are two examples of torsion bars being used to eliminate the transverse leaf spring. Both cars utilize a split wishbone to locate the axle relative to the chassis. In the case of the car still under construction, a beam axle has been retained, but cut in two and hinged in the center to work in conjunction with an upper A arm structure which is also a part of a torsion bar assembly. This is far too complicated for most home builders to pull off.

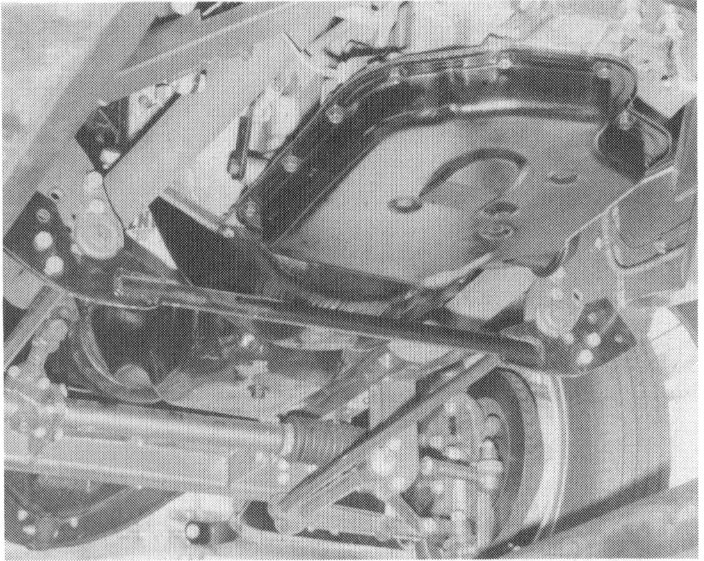

(Above and Below) Here are a couple of fine examples of Jag suspensions grafted to early Ford frames. Again, this is best left for the professional car builder — and some of them don't get it correct either!

VEGA OR H-CAR CROSS STEERING

Although Ford used cross steering on the Model T and again from '35 thru '48, only in the past few years has this style of steering become a widely accepted form of steering for rods based on '28 to '34 Fords. Recent interest is due primarily to the compact size, strength and availability of the steering box found in 1971—1981 GM H cars. This steering box is erroneously referred to as a Vega box when in fact it should be known as "H-car" for it is found not only in Vegas, but also in Monzas, Starfires, Skyhawks, Astres and Sunbirds — and in GM manufacturing nomenclature, these are all H-cars.

You should also know there are two different H-car steering boxes. They look the same, but they are different on the outside (one is larger by about 20%) and they are different on the inside (different size parts that do not interchange). When the Vega was first introduced, it was equipped with what we'll call the standard steering box. This is the smaller of the two units and has a gear ratio of 20.9:1. A heavy duty box was introduced with the Cosworth Vega and then installed on an increasing number of H cars over the years. By the time the H car production was halted in 1981, the heavy duty box had become standard. The gear ratio of this box is 16:1 — making it considerably quicker than the smaller box.

There are several ways of identifying the two H-car steering boxes. One simple way is to measure from the leading edge of the large locknut at the front of the steering box to the most rearward portion of the iron steering box housing. The standard box will measure roughly 5¼-inches there, while the heavy duty version will

(Above and Below) When a Vega steering box is utilized in a cross steering car, the box must be placed far forward on the frame, whereas the Mustang box gets mounted almost directly below the firewall. In both cases, clearance with the engine and the brake pedal must be taken into consideration. In both of these examples note the use of an industrial universal joint to join steering box to the column.

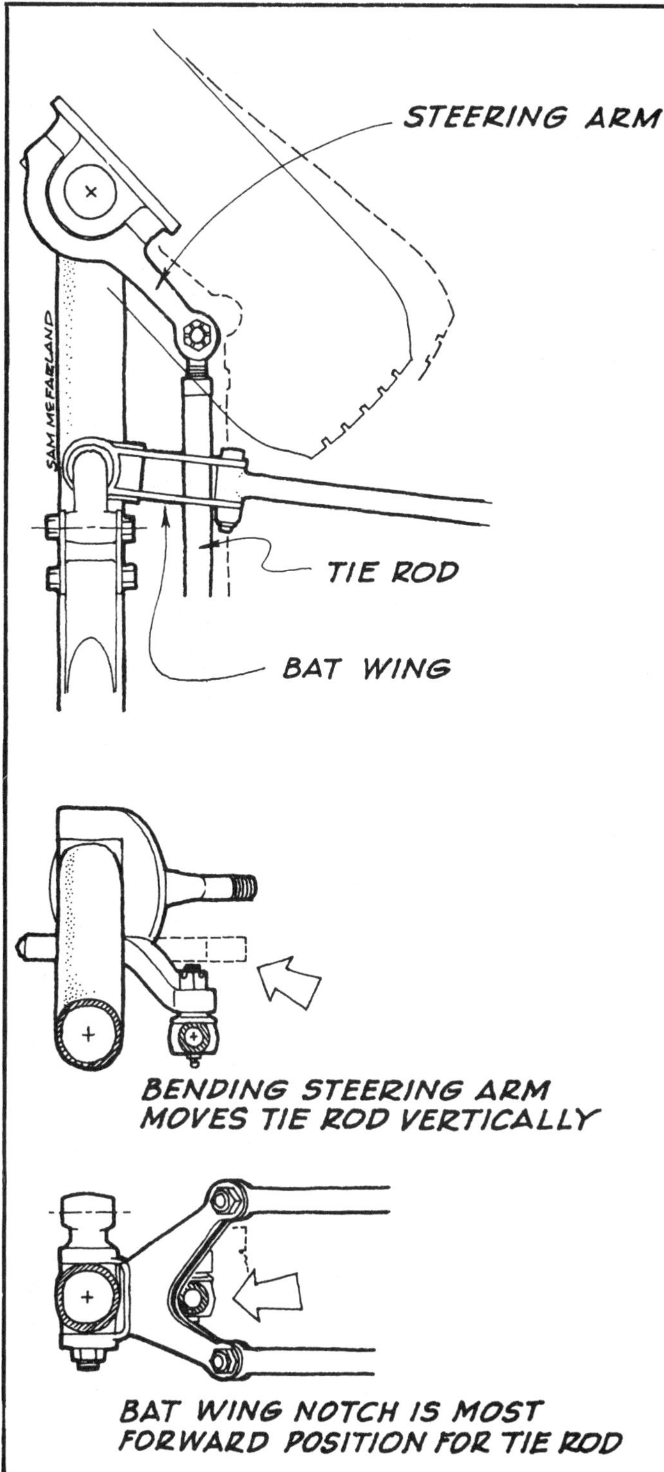

Here's a good example of putting the tie-rod in the deepest part of the batwing notch to reduce the turning radius of the car. This is done by heating the steering arms and bending them.

be closer to 7⅛-inches. If you are contemplating building a fenderless repro rod, you might like to know that with H-car cross steering, the pitman arm and drag link are neatly hidden beneath the front of the car which gives a very clean, uncluttered appearance. You also might like to know that in many installations, the larger of the two H-car boxes cannot be used due to clearance problems.

In cross steering, the steering box mounts far forward on the inside of the left frame rail. The drag link runs laterally across the chassis where it connects to the steering arm of the right spindle. The pitman arm pivots side to side moving the drag link laterally to steer the car.

Cross steering is not affected by the type of radius rods used to hold the axle. As we learned by the illustrations concerning the Mustang/Cougar steering, the use of a four bar or a wishbone with a particular steering set-up can mean the difference between a good or bad handling car. This is not the case with cross steering. Front movement as controlled by the radius rods which locate the axle front-to-rear is not relative to the steering movement because the drag link is mounted laterally. This leads to the assumption that ill-handling characteristics such as bumpsteer do not exist with cross steering. Not so.

The steering or directional change of the car is controlled by lateral movement of the drag link. In other words, when the steering wheel is turned, the drag link will move laterally in relation to the axle/spindle assembly causing the spindles to turn and the car to change direction. If the drag link moves in relation to the axle assembly without the steering wheel being moved the spindles would still turn and the car would still change direction. This can happen due to the fact an early Ford style transverse spring is free to move from side to side on the shackles. This lateral movement not only occurs from bumps and cornering but is a continuous action resulting from any suspension movement. There are cases where lateral movement is restricted by stiff working shackles or shackles that don't swing at all because the main spring leaf is too short.

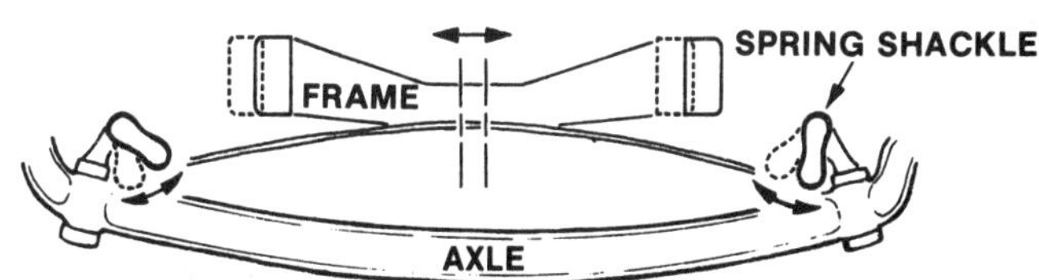

Drawing courtesy of Pete & Jake's

Oversteering is also a problem with cross steering. This phenomena is most obvious during hard, fast cornering. Take a left turn for example. The steering wheel is turned to the left an amount adequate to take the turn. The drag link moves to the right which turns the spindles a corresponding amount to the left. As the car responds going into the turn, centrifugal force causes the weight of the car to shift to the right moving the frame to the right on the shackles. The steering gear and drag link being relative to the frame also move to the right. This additional drag link movement then causes the spindles to turn more than desired. This is oversteer — with the

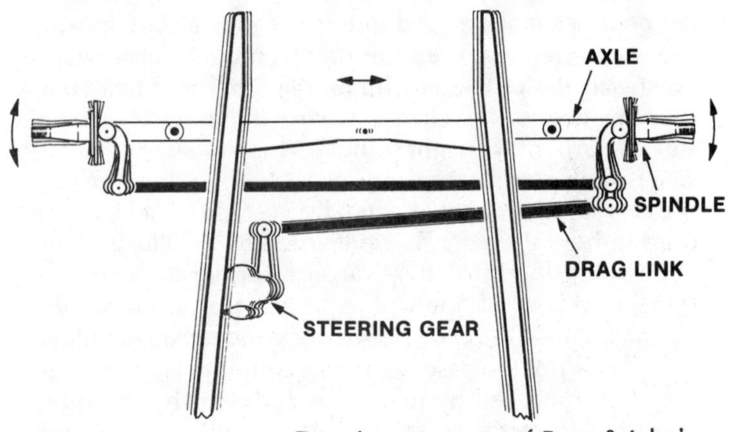

Drawing courtesy of Pete & Jake's

ultimate manifestation being spin-out.

The answer is to eliminate or control the lateral movement between the frame and the axle. This is done by installing a Panhard rod, or sway bar, which is a lateral link connecting frame and axle.

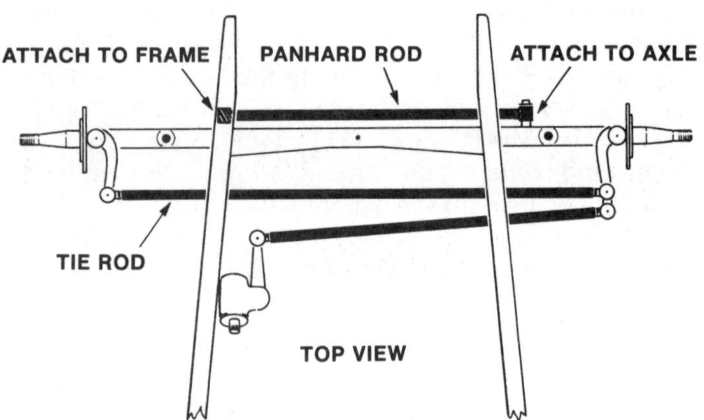

Drawings courtesy of Pete & Jake's

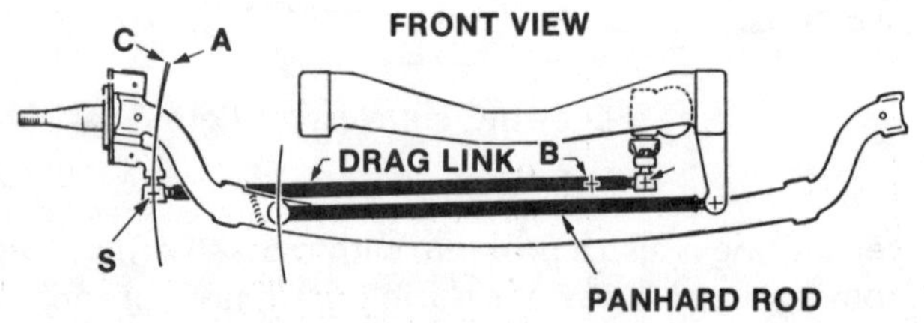

Lateral control of the axle must be relative to the drag link for correct steering geometry during vertical suspension movement. Because the Panhard rod pivots at both ends, like the radius rods of a four bar suspension, correct geometry is based on the parallelogram in that the drag link and Panhard rod should be parallel and as close to the same length as possible. As part of the axle/spindle assembly, the steering bar on the spindle arm (point S) travels an arc (A) equal to the length of the Panhard rod, centered at a point (B) parallel to the Panhard rod. Point S also travels an arc (C) centered at the other end of the drag link (point D). If the drag link and Panhard rod are parallel the arcs will be very close within the limited amount of vertical travel and steering will not be affected. If they are not parallel, the arcs will not be as close and the resulting bumpsteer will depend on the degree of angle difference between the two.

Clearance problems need to be taken care of as the project moves along. Don't let them build up and become intolerable. It is important to keep in mind that sufficient space must be provided for other items of hardware and that the car will need to be serviced from time to time.

STEERING COLUMNS

There are a number of factors to consider in selecting a steering column for a repro rod. If the car is to be equipped with an automatic transmission, do you want the shifter to be a part of the steering column or do you want the shifter on the floor? If you want the shifter on the steering column, do you want the column to tilt?. Do you want factory cruise control? Now we're narrowing the field in a hurry. The problem seems to be that there are plenty of columns around that tilt, have provisions for cruise control and an automatic transmission shifter, but there are few columns which have the shift quadrant indicator on the column as opposed to being part of the dash. If you have plenty of skill, time and patience, the shift quadrant indicator can be made a part of a repro rod dash, but this is easier said than done.

There are at least two columns that meet the above criteria: late GM vans and '68-'69 Thunderbirds. Undoubtedly there are others that a careful search through a wrecking yard will turn up.

The shifter on the column is desirable because it yields considerably more floor space than allowed by a floor shifter. This is important when three people are in the car. The tilt wheel and cruise control are real creature comforts if you plan to do a lot of long distance driving.

If you feel you can do without any of the above embellishments to a plain jane steering column, the choice of columns will be greatly expanded. Regardless of the column chosen, it should be anchored firmly to the dash panel or dash panel reinforcement if a glass dash is being used in a glass car and to the floorboard. A firmly anchored steering column is essential for precise steering; the plan is for you to steer the car and not have the car steer you.

(Above and Below) Regardless of the car or the column used, the column must be firmly anchored to the body at the dash and to the firewall or floorboard.

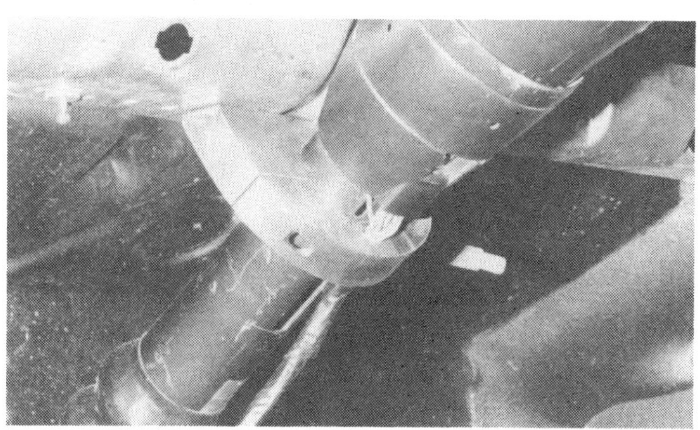

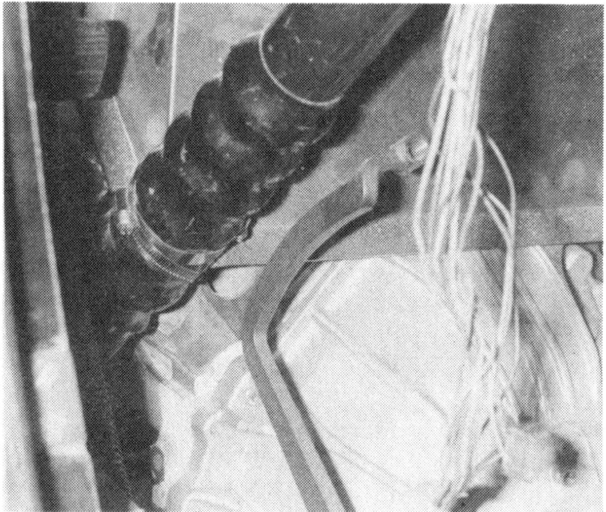

Very early in the mock-up stages of installing the steering column make sure the brake pedal assembly is in place. Don't be surprised if you have to heat and bend the pedal arm to place the pedal away from the column.

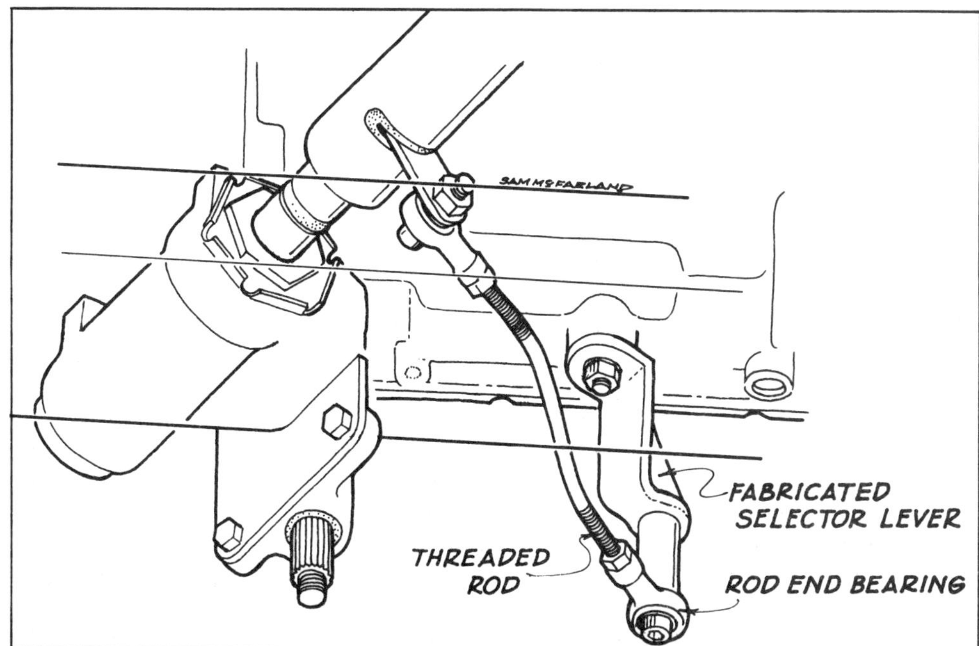

One of the important items to keep in mind when selecting and installing the steering column is how you will solve the trans shifter hardware problem. This is a good example of how it can be done using ball joint rod ends at each end of a threaded rod. Once the final adjustments are made, the rod ends can be locked into position with jam nuts.

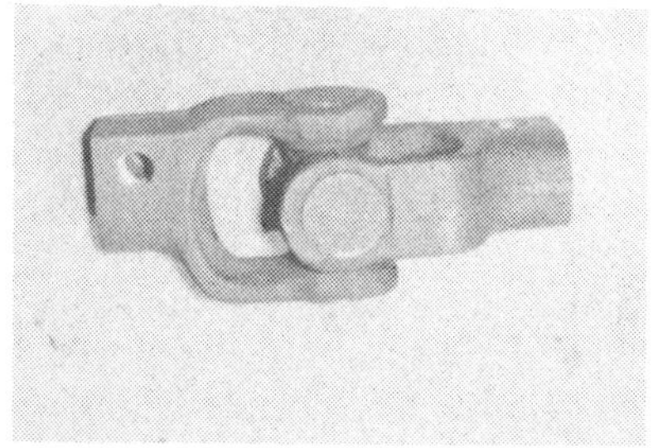

Don't scrimp on the quality of the U-joint you select to join steering box to column. This is an excellent item used on Olds Toronados. Each coupler is splined and drilled and tapped for a set screw.

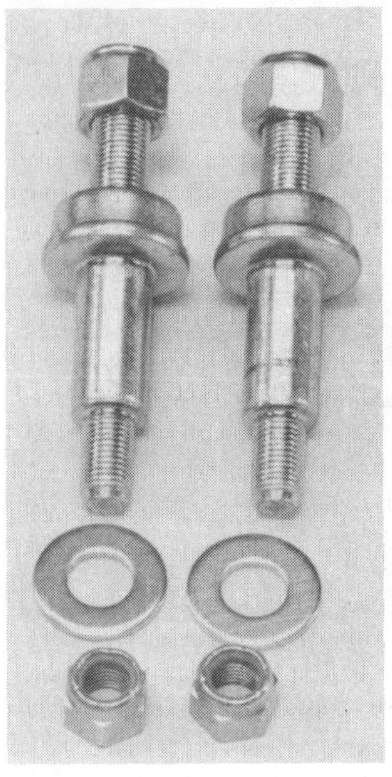

Shock studs such as this are available from Pete & Jake's and in conjunction with a frame mounted bracket can be used as an upper shock mount either front or rear.

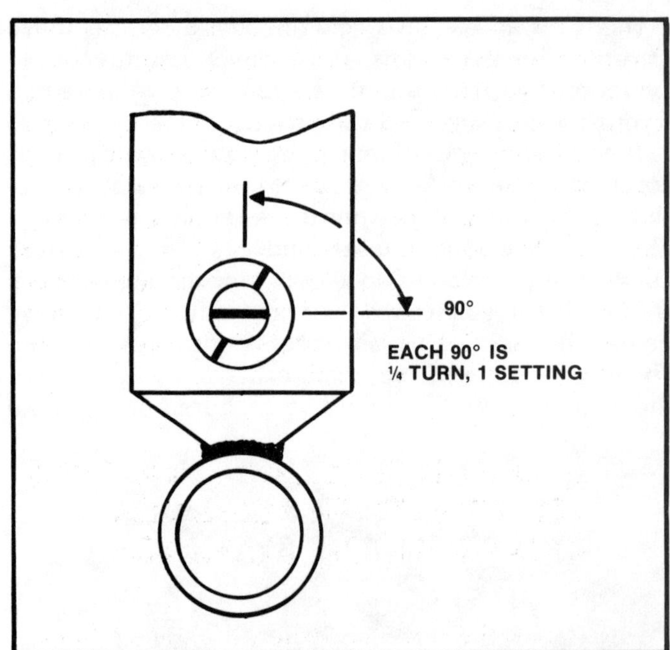

SHOCKS

We've come a long way in the past twenty years of rod building in reaching the state-of-the-art shock. Friction and frame-mounted lever shocks have fallen by the wayside for all but the die-hard nostalgia freaks. We've gone through a period where the "hot tip" was to use worn out taxi shocks (because they were soft). Currently the state-of-the-art is a fully-adjustable shock in either plain (for the front suspension) or coil wrapped for the rear. By fully-adjustable we mean the dampening rate can be changed with the edge of a coin or a screwdriver in a slot. On a coil-over shock, ride height of the spring can also be adjusted as much as two inches by moving a threaded collar beneath the spring. This is in addition to the previously mentioned dampening adjustment. As this book goes to press the only shock fitting the above description is the Aldan shock. Koni and Carrera are also adjustable to a certain extent — but not with the ease of the Aldan. This situation will most likely change in time. Pete & Jake's and Magoo are probably the best places in the country to obtain current information on shocks.

Many rodders misread what a contemporary street rod needs in the way of shocks and suspension. We're involved with a light vehicle outfitted with radial tires running primarily on very smooth surfaces. Hard shocks and stiff springs are simply not needed or wanted.

When shipped, all Aldan shocks are set at the lowest setting. This means the valving control screw is rotated in a counter-clockwise direction until it stops as shown in the illustration. To increase the valving and make the shocks stiffer, the control screw is rotated in a clockwise direction. There are 15 self-locking positions, each one being felt in the form of a click every ¼-turn. The setting on the first four positions is not clearly distinguished — just keep in mind each quarter turn represents one adjustment.

In mounting shocks, keep in mind that for full working use, a mounting angle of 20 degrees is minimum, 30 degrees is ideal and 45 degrees is the maximum angle that a shock should ever be mounted. When the shock is mounted it is imperative that the top and bottom mounting brackets be perfectly parallel to each other and the shock mounts with no binding.

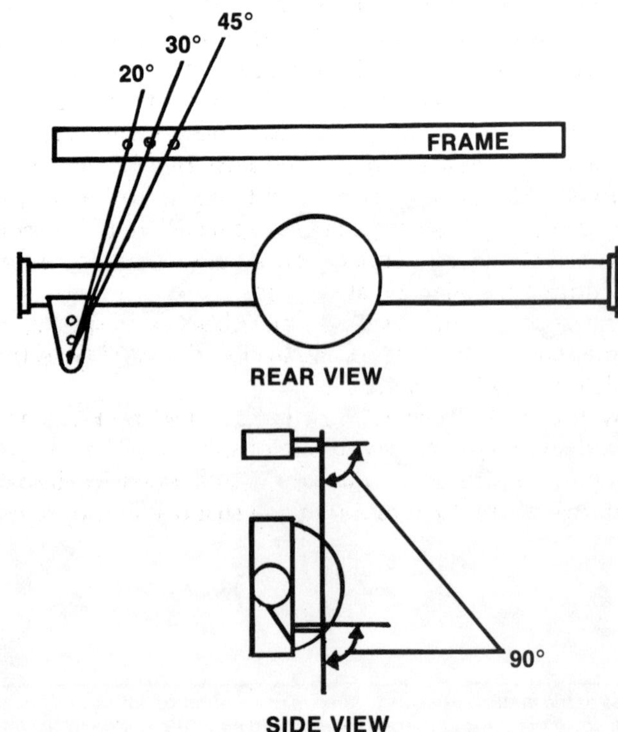

Make sure the shocks do not "bottom out" or fully collapse before the suspension bottoms out on some sort of bump stop. If a shock bottoms out before the suspension does, damage to the shocks or the mounting brackets will occur. To check for potential bottoming, it is suggested that you bottom out the suspension, frame against rear axle housing, with the rear spring removed; then, select the position that will mount the shock in an "almost" fully collapsed position. If there are rubber snubbers between the frame and the rear axle, be sure to allow for the fact that these snubbers will collapse before the suspension is fully bottomed out.

ROD END AND SPHERICAL BEARINGS VERSUS RUBBER BUSHINGS AND URETHANE BUSHINGS

Sometime in the nineteen sixties, the use of "trick," "aircraft," "surplus," or "race car" rod ends became very popular among those building "rods." The classic case of "monkey see, monkey do" escalated to the level of "non-think" among some rod builders and the result was some dangerous, expensive rods rolling down the highways.

In his excellent book PREPARE TO WIN, Carroll Smith has this to say about rod end and spherical bearings:

"Suspension linkage bearings of the rod-end and spherical variety present a problem in that there are a lot of them; they do wear out; good ones are expensive and bad ones are an abomination."

"Trick" rod ends and spherical bearings come from the aircraft industry. Generally speaking, they are lightweight and precise. They were designed to perform a specific task within an air frame. There is a considerable difference between this environment and that beneath the fenders of a rod driven on public roads. Modern race cars and airplanes are inspected at frequent regular intervals. Normally, street rods are not. Dirt and corrosion of any kind takes its toll on any precision part which is precisely why "trick" rod ends should stay on aircraft and race cars and not be installed on street rods.

The second vote in favor of rubber bushings is as elementary as the first — the overall driveability of the vehicle is infinitely better.

By the mid-seventies most serious rod builders had drifted away from blanket usage of spherical rod ends in street rod suspension and were using press-in rubber bushings of the type found in the susupension of many domestic cars and trucks of the era. These pieces were readily available, proven in usage and greatly improved the overall driveability of the vehicle due to the simple fact that every vibration, and bump encountered by the suspension was not transferred steel to steel into the frame, the body, and the seat of your pants. For about five years, most of the respected rod building and component shops such as Challenger, Pete and Jake's, Kugels and Magoo's hung front suspension, Panhard bars and shocks in rubber bushings encased in tubing and retained conventional tie rod ends for the steering mechanism.

In the late seventies Pete and Jake's pioneered the use of urethane bushings in the four bar assemblies. These bushings (and those of other manufacturers to follow) are direct replacements for those four bar assemblies that previously had been outfitted with the press-in rubber GM bushings. At this writing, the urethane bushings in the front suspension are the best compromise available. They eliminate the considerable compliance and "squirm" associated with the widely popular rubber bushings and they offer the precise control which can be associated with metal spherical bearings.

Hopefully, the position is clear about not using spherical bearings — in terms of safety and long life that is paramount in the construction of a 1980's repro rod. While we're standing on the soap box, you should know we feel the same about using aluminum bolts in the running gear. This is not intended to be a blanket condemnation of aluminum fastners, but for the average rod builder who winds up with some aluminum bolts of unknown quality, the situation is an extended form of Russian roulette. Use a quality steel item that was built to do the job.

You should also know that in terms of ride quality there is one step up from the commercially available Nylon, Urethane, Polyurethane, etc., bushings. All of the above is good, but machined Teflon is better. Unfortunately, no one has figured out how to injection mold Teflon — it must be machined to produce a part like a rod end bushing. due to the difference in the manufacturing technique, the Teflon bushings are much costlier than the other forms of "plastics"; thus mass merchandisers have opted not to produce the Teflon bushings at this writing. Teflon has an oily, self-lubricating quality about it that other "plastics" do not have and the difference is evident in a street rod. As this is being written we know of only one source for machined Teflon busings — Magoo's.

GLOSSARY

Anti-roll Bar — A spring medium, usually a bar or tube anchored to the frame and linked to either the front or rear suspension members to provide roll resistance and designed to function primarily during body roll. The bar is often mistakenly referred to as a sway bar or anti-sway bar. This anti-roll bar can be helpful with a coil-over rear suspension street rod which gives a soft ride but tends to have increased roll in the chassis.

Bat Wings — These are the brackets which anchor the front attachment points of a four bar assembly to a front axle. Bat wings must be matched to the width of the axle — either 2 or 2¼-inch. Normally they are sold only with the rest of the four bar assembly.

Bigs and Littles — This is a phrase coined by Gray Baskerville of Hot Rod Magazine to describe the "large tire in the rear, small tire in the front" syndrome which has become so very popular in giving an increasing number of street rods "the California rake."

Brake Proportioning — The equal or unequal distribution of braking force among the four wheels.

Bump Steer — Changes in the toe (in or out) settings during vertical suspension movements resulting in the car steering itself. Obviously this can be hazardous to your health!

Bump Stop — A device which limits the bump travel of a suspension system so the suspension or the shock absorber will not bottom out. Bump stops can either be elastic (rubber) or solid.

Camber — The inward or outward tilt of a wheel at the top relative to the vertical at the centerline of the wheel in the lateral plane. Zero camber is true vertical. Negative camber is the tilt of the top of the wheel toward the center of the vehicle. Positive camber is the tilt of the top of the wheel away from the center of the vehicle.

Camber Angle — The angle measured in degrees -- positve or negative -- between true vertical and the vertical centerline of the wheel. Positive camber is when the wheel leans outward at the top and negative when it leans inward.

Camber Change — The increase or decrease in camber angle during suspension movement.

Caster — The inclination from vertical of the steering axis in the longitudinal plane.

Caster Angle — The angle in degrees positive or negative between true vertical and the steering axis in the longitudinal plane. Zero caster is a steering axis that is vertical. Positive caster is when the steering axis is inclined toward the rear of the vehicle in the side view (top of the king pin further to the rear than the front).

Chassis Rake — The practice of lowering the front (or raising the rear) of a chassis so the bottom of the frame is not parallel with the ground.

Coil-Over Suspension — A suspension system where the shock absorber is mounted inside a coil spring resting on adjustable height plates mounted to the shock with the entire unit having common mounts on the chassis and suspension linkage.

Cross Steering — The most popular type of steering for use on a fenderless street rod due to the fact the drag link and pitman arm are hidden under the car. Essentially the passenger side of the rod is steering first and the tie rod attached to the right side steering arm transmits the force to the driver's (left) side. Any cross steering, solid axle, transverse leaf spring-equpped street rod should be equipped with a Panhard bar (rod). See text for reasons why.

Curb Weight — The net weight of a vehicle with fuel, oil and water but no driver.

Drag Link — A steering linkage used on cars with beam (solid) front axles that runs from the pitman arm on the steering box to a steering arm which is at right angle to the wheel plane.

Dual Braking System — A braking system utilizing two master cylinders -- one for the front wheel brakes and the other for the rear wheel brakes.

Four Bar (Point) Suspension — A suspension system on a beam axle using four longitudinal control arms attached to the axle at one end and the chassis at the other to locate and control suspension motion.

Free Length of Spring — The length of a coil spring measured from end to end with no load on the spring.

Independent Rear Suspension — A rear suspension system where the differential is mounted to the chassis and each rear wheel is allowed to move independently. Jaguar and Corvette rear suspensions are the most popular independent rear suspension units being used on street rods.

King Pin — The pin or bolt about which the steered front wheels pivot in the vertical plane.

King Pin Inclination — The angle from true vertical that the king pin rotates about in the lateral plane.

Magoo (Nickname for Dick Megugorac) — One of the top two or three rod builders in the country. A "Magoo Car" is one which is well engineered and flawless in detail -- regardless of who built it.

M.I.G. Welding (Metal Inert Gas Welding) — An electric welding process where the filler material is wire-fed through the torch from an automatic feed roll. The wire is the electrode which is surrounded by an inert gas, usually carbon dioxide or argon, which shields the weld area from oxidation.

Mild Steel — Low carbon steel or any carbon steel where the carbon content range is between .08 - .35 percent. Usually designated with a number such as 1010, 1015, 1020, etc. This is excellent material for fabricating small pieces on a street rod.

Neutral Steer — A vehicle is neutral steer at a given trim if the ratio of the steering wheel angle gradient to the overall steering ratio equals the Ackerman steer angle gradient.

NSRA — The National Street Rod Association. A nationwide organization dedicated to the sport of street rodding.

Oversteer — is felt by the driver as a tail-out yaw angle, requiring a steering correction in the direction opposite to the direction of travel; ultimately the vehicle will spin. Oversteer occurs when the rear tire slip angles exceed the front tire slip angles.

Panhard Bar (Rod) — A bar or rod used to control lateral movement of a body or chassis relative to the axle. A lateral control linkage. One end is attached to the chassis and the other end to the axle.

Pitman Arm — A steering system linkage mounted to a steering box shaft on one end and to a drag link or steering cross link at the other. A pitman arm is used to change the rotary motion of the steering gear into linear motion.

Reversed Eye — The act of heating, bending and repositioning the shackle eye of a leaf spring from the bottom to the top of the spring thereby lowering the car. This is most commonly found on transverse leaf spring street rods.

Ride Height — The height from the ground level to specific points on the chassis in the static state with the vehicle weighted as it would normally be driven.

Ride Height of Spring — The height of a coil spring installed in a vehicle weighted to normal driving weight, measured while the vehicle is at rest.

Rim Width — The distance between the inside surfaces or the rim flanges.

Shackle — A moveable connecting link found at the end of a leaf spring. A transverse leaf spring will have a shackle at both ends.

Shackle Bushing — A bushing fitting between the pin of the shackle and the eyelet of the leaf spring. Current street rod useage dictates the use of Nylon, Teflon or some other "plastic" for optimum ride.

Shock Absorber — This is a misnomer. Shock absorbers do not absorb shocks. Shock absorbers dampen the natural oscillations of a spring.

Shock Absorber Collapsed Length — The distance between mounting point centers when the device is fully compressed.

Shock Absorber Control Ratio — The comparison of dampening forces in "rebound" and "compression." This is expressed as a percent of total control.

Shock Absorber Extended Length — The distance between mounting point centers when the shock is fully extended.

Shock Absorber Stroke — The difference in extended and collapsed length of a shock absorber.

Spindle — A rod, pin or shaft that rotates, or serves as an axis for revolving parts.

Spring Arch — The curvature of a leaf spring; a negative arch is called de-arch.

Spring Perch — On a street rod this is normally a pin containing an eyelet with the pin anchoring through the front axle and the eyelet becoming the mounting point for the transverse leaf spring — via the shackle.

Steering Arm — An arm, either integral with, or attached to, a wheel spindle, connected to a tie rod which provides steering inputs.

Steering Ratio — The ratio of the gears within a steering box.

Suspension Geometry — Wheel motions relative to the chassis determined by the arrangement of suspension components.

Tie-Rod — A linkage in the steering system that connects the steering arms of the left and right front wheels on a beam axle car. All tie-rods are adjustable in length to accomodate toe settings.

Tie-Rod Ends — A special ball joint with a threaded shaft which screws into the end of a tie rod.

T.I.G. Welding — Tungsten Inert Gas Welding. A type of electric welding process using a non-consumable electrode in the torch, externally, hand-fed filler rod and argon gas to shield the weld area from oxidation. Used on alloy steels, stainless steel, aluminum and other exotic metals. Commonly called heliarc.

Transverse — Lying, situated, or placed across; crossing from side to side; crosswise, lateral.

Tread Width — The distance between the extreme edges of road contact at a specified load and pressure measured parallel to the Y axis at zero slip angle and zero inclination angle.

Toe-In — The difference in distance between the front and rear measurements of tires on the same axle, in the center of the tread surface at spindle height where the front measurement is less.

Toe-Out — The difference in distance between the front and rear measurements of the tires on the same axle in the horizontal plane measured on the center of the tread surface at spindle height, where the front measurement is greater than the rear.

Toe-Steer — A steering effect due to toe changes, caused by suspension or steering linkage geometry during vertical wheel deflections.

Understeer — is felt by the driver as a nose out (to the outside of the turn) yaw angle requiring a steering correction in the same direction as the direction of travel; ultimately the vehicle will continue in a straight line. The front tire slip angles exceed the rear tire slip angle.

Watt's Linkage — A lateral locating linkage using a center pivoted bellcrank and two parallel links running one from each end of the bellcrank in opposite directions to the chassis or frame of a vehicle allowing vertical or rotational travel of the axle or axle housing without lateral displacement.

Rear Axle and Suspension

Every open driveshaft, live axle known to man has, at one time or another, been used under a street rod in the last twenty five years. Dozens of articles have been written on how to hang them, how to narrow them and how to change the bolt pattern. Although there was obviously some justification for doing all of this at the time, there is little or no reason for making a lot of work for yourself when selecting a rear axle for a modern repro rod.

The perfect width axle for a '32-'34 Ford is 57½-inches measuring from outside of drum to outside of drum. A model A will also accept a 57½-inch axle, but attention must be paid to the tire and wheel selection if you plan to shove all of the tire under the rear fender. The perfect rear axle width for a full fendered A is 56½-inches. Ford passenger car axles found in '57-'59 models measure 57½-inches. Ford Mustang and Maverick rear axles measure 56½-inches. There are two versions of the latter axle. Those Mustangs and Mavericks built with V8 engines had a bolt pattern on the wheels of five lug on a 4½-inch circle. Six cylinder cars had a bolt pattern on the wheels of four lug on a 4-inch circle. The axle tubes on the six cylinder cars are necked down between the center section and the backing plate while the axle tubes on the V8-equipped cars are straight.

There is only one fly in the ointment when putting either of the Ford axles mentioned under a repro rod. The center section is offset to the left side. This simply means the center section will not be centered between the frame rails. Because the rear axle assembly is a visual part of the rear end of many a repro rod, this off-set might not be pleasing to your eye. The right side of the axle assembly can be shortened two inches and the center section can then be centered. This should only be done with the 57½-inch axle because removing two inches from the Maverick/Mustang axle simply makes the entire assembly too narrow for anything but an A with very skinny tires and wheels, or a T.

Unless you do something really weird to a '32-'34 repro rod, very little of the rear axle will be visible to the casual observer so the concern about the offset in terms of visual impact should be taken into account only when building a Repro Model A. In building the A roadster as part of the research for writing this book, we used the V8 Mustang rear axle and did not have the right side shortened. The axle is visible and is part of the visual impact of the rear of the car. We simply painted the axle housing and forgot about it.

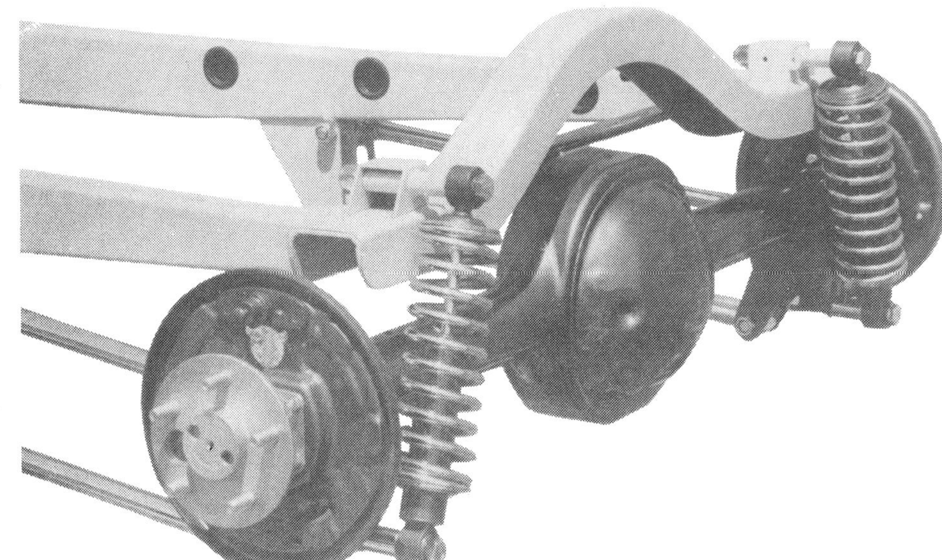

This is a state-of-the-art coil-over, four bar, Mustang rear axle, hot rod rear suspension.

Rear Suspension

There are several ways to suspend the rear of a repro rod. Over the years the drift has been away from the stock type transverse leaf. There are a couple of reasons for this. In the first place, leaf springs do not give as good a ride as coil or coil-over-shock suspensions. And, it is difficult, if not impossible, to get a rod as low as most builders desire if a transverse leaf spring assembly is used.

Because hardware is available from a great number of sources which will allow you to install transverse leaf springs on a late axle, we will comment further on this set-up before moving on to more modern and practical rear suspension assemblies. The ride of a transverse leaf assembly can be improved if the spring package is reduced to approximately eight leaves instead of the normal stock assembly of twelve. The ends of each leaf should all be rounded with a grinder to keep them from digging into the next lower spring. Strips of polypropylene or Teflon should also be installed between all leaves to ease the sliding action between the leaves.

As the instruction sheet from Pete and Jake's point out, there is more than meets the eye when putting a late rear axle in a Model A while retaining the transverse spring assembly. If for some reason you cannot hold the 49½-inch dimension between the spring mount tubes, the spring will have to be narrowed by a competent spring shop. Good luck. This is exactly the sort of "simple" task that drives the novice rod builder up the side of the wall. First you have to find that competent spring shop, then you have to explain what you want done. Of course this will not be the same shop that you select to weld the brackets to the axle housing. Now we have another cook stirring the stew. Note the admonition in the instructions that adequate clearance must be left between the spring

bracket and the brake line coming from the wheel cylinder. Note, also, that once you have all of this done you still have to make provisions for mounting shocks in the rear. Kits are available for this however, which means you get to visit the welder again. If a transverse leaf spring is adapted to a late model rear axle for use under a repro rod, then traction bars (also called ladder bars) must also be installed. Failure to do this will result in the imminent failure of one or more of the rear suspension points in addition to one of the universal joints.

Why do we need traction bars when a transverse leaf is used with a late rear axle under a repro rod? Stock, early Fords had no such arrangement. But, they did have a

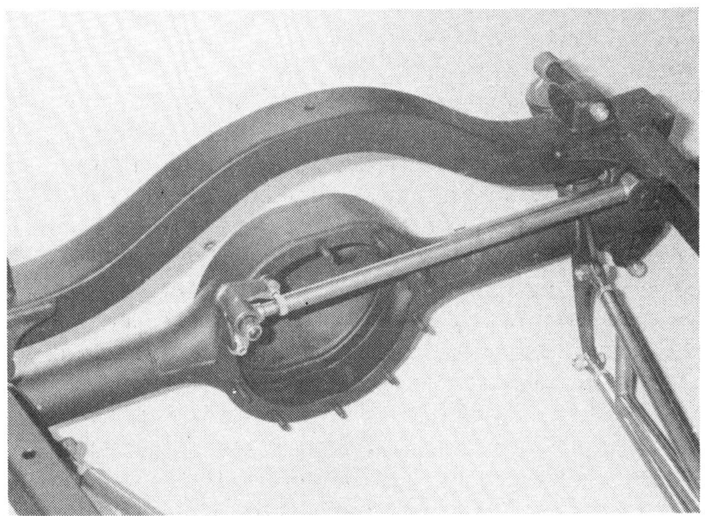

A Panhard rod is the most logical piece of hardware to limit lateral movement of the rear axle when coil over suspension is used.

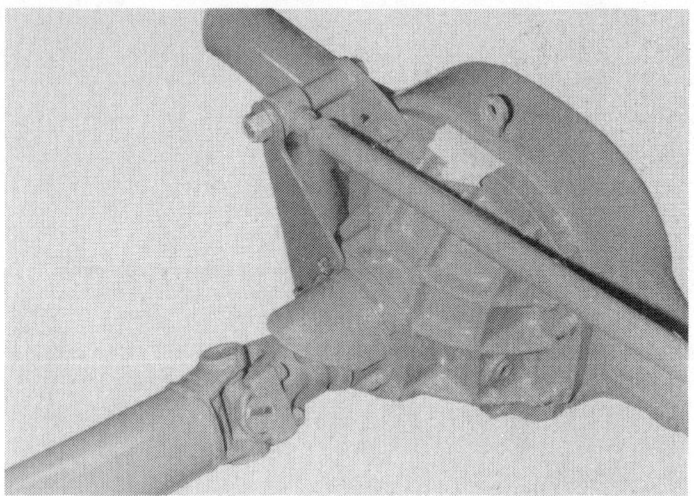

Don't take the mounting of the Panhard bar lightly. Both ends of the bar should be captured on both sides. This bar and the attending brackets must withstand tremendous strain when the car is cornering.

rather undramatic looking piece of steel called torque tube. The torque tube surrounded the driveshaft and was attached to the center section of the rear axle and to the rear of the transmission in order to limit fore and aft movement of the rear axle. Traction bar kits — which take the place of the old torque tube — are available from a number of sources, so it's back to the welder.

Fortunately for all of us (and the welder), no anti-sway bar or Panhard rod is needed with a transverse rear spring assembly. The spring itself limits lateral movement.

JAG SUSPENSION

"Pussy foot" Jag suspension became the "in" thing for the street rod building set in the early seventies. Cars equipped with Jag underpinnings got a lot of play in the enthusiast magazines and a lot of attention at rod runs. The reason for this focus is simple — the setup looks neat. But there is more to achieving this part of "neat" than meets the eye.

In the first place you actually pay for Jaguar suspension three times. The first time is when you cart it home from the wrecking yard. The second time is when you pay a lot of extra money to get all of the extra bracketry mounted on your repro chassis. Even if you are a pretty fair fabricator and welder, think twice before plunging into a project which demands that all suspension points be critically located for optimum handling. The third time you pay for

There are several sources for weld-on rear coil-over shock brackets that also double as bumper bracket supports on a Model A.

a Jaguar suspension is when you pick up all the pieces at the chrome shop. When is the last time you saw non-chrome Jag suspension under a rod?

Other than looking neat, the argument for Jag suspension is that it gives a ride far superior to that provided by straight axles. That may be true, but with today's radial tires and smooth streets and highways a solid axle car gives a good enough ride for my skinny, sensitive rear end.

If you have a Jag-suspended car or have your heart set on building one, please do not be offended by our remarks. Jag suspension — properly done — is safe, smooth, looks neat and is expensive.

Unless you are very knowledgable on suspensions or have access to someone who is, I don't feel a first time rod builder has any business messing with it. If you do not have Jag suspension under your rod, you should never feel you are a second class citizen or rodder because of it.

This is a well-thought-out installation of a modified Corvette rear suspension in a Model A. This is definitely beyond the skills capability of most rod builders.

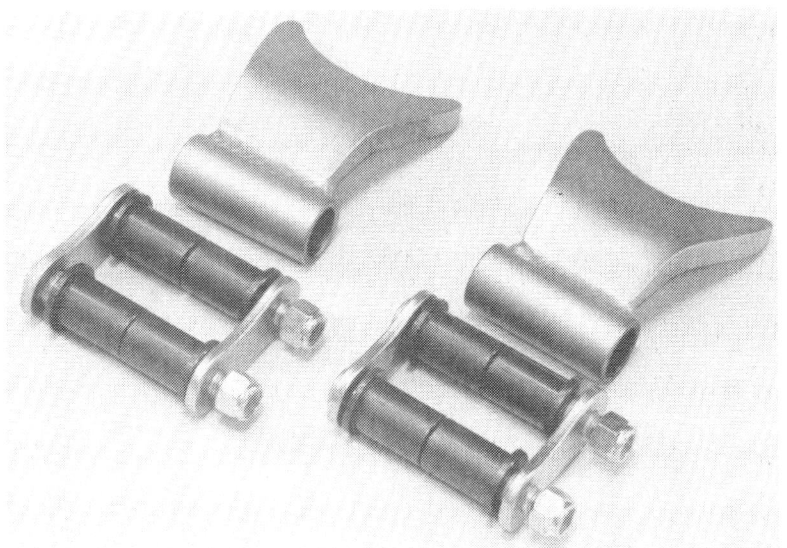

If you want to install a late rear axle under a transverse leaf spring in the rear, weld-on brackets and shackles are available from a number of sources. Measure the spring width before ordering the shackles.

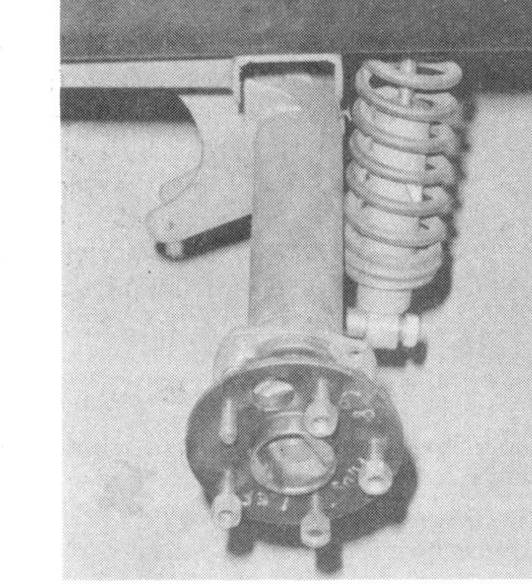

(Above Right) It's not a bad idea to have the body on hand before locating the rear suspension hardware beneath the frame. You'd be surprised how many street rods are running around without having the wheel centered in the wheelwell.

(Right) When setting up the rear suspension, use a magnetic base plumb and level to get the forward axle housing flange parallel with the rear face of the transmission. Failure to do this can cause strange driveshaft harmonics.

Drivetrain

BUYING AN ENGINE FOR THE REPRO ROD

Buying an engine for your repro rod should be carried out just like your approach to building the entire car. You must consider what is practical for you. Cash, time and skill all must be considered if you are to make a wise judgement in what you finally buy. Take your time and think it through, then do a lot of shopping. Engines for repro rods can be divided into three general categories — new, used, and rebuilt or remanufactured. There are advantages and disadvantages in buying an engine in any of the above categories. Don't overlook that and do not overlook the very simple fact you must determine what is practical for you in your situation.

New Engines

Obviously the advantage of buying a new engine is that it is new. A new engine will have been assembled by a factory that builds hundreds of thousands of the same engine every year. With regular oil and filter changes and tuneups, most domestic engines will give upwards of 100,000 miles of good service before needing a rebuild. That's a lot of miles to put on a street rod. Those who neglect engines in various ways have their own justifications for doing so and are thus not entitled to 100,000 maintenance free miles of street rodding.

New engines are available two ways — complete or in pieces. In either case you go to the dealership parts counter to get the hardware. Although you may want to buy a complete engine, something will be missing — a starter, an air cleaner, an alternator, or something else

Just about the time you think you've seen everything you find out you haven't. This Ferrari V12 has been mated to a Doug Nash 5-speed transmission for use in a street rod. Undertakings like this are left for those with a lot of talent or money — or both!

(Above and Right) Although the days of the V8 are severely limited, they will continue to be a very popular engine for street rod use for many years to come. Their usage increases the complexity of building a modern rod. In this case the firewall had to be recessed to get the engine under a stock length hood. This shoves the transmission back into the floorboard which means more fabricating to get the floor to clear the transmission.

In sharp contrast to recessed firewalls and reworked floorboards, here's a turbocharged Buick V6 installed in a '29 hiboy without altering firewall or floor.

you want for your car. In most cases the rodder who is knowledgeable enough to pull this off will be lacing new and used parts together with a scattering of aftermarket parts thrown in for good measure. The advantage to building an engine in this manner is in putting together exactly what you want. If the rodder is blessed with talent, patience and plenty of common sense, the results can be most rewarding.

The drawbacks to this approach are many for the novice rodder. Time, machining costs, and the possibility of putting together combinations which don't work out in terms of reliability or power all take their toll.

Do not pass over complete (or nearly so) engines lightly. Visit the local dealership handling the engine you desire. Tell them specifically what you have in mind. Find out what the engine costs, and exactly what you get for the money. When they are talking about a complete engine, find out if this assembly includes the essential items of hardware such as carburetor, intake manifold, distributor, wiring, starter, alternator, water pump and fan. Make a list of the questions and take pen and pencil with you so you can write down the answers.

which is basically a bolt-on item. Don't lose sight of the fact that somehow or another you will have to come up with the hardware it takes to make the engine complete. Buying all of this new can be expensive. But when you weigh all of the factors involved you may very well find this as the least expensive way *for you* in the long run.

Buying a new engine in pieces is obviously not for everyone, but you should know about it. This plan of attack involves going to the dealership parts counter and buying select pieces of hardware to build just the engine

Used Engines

The great majority of all engines used in street rods start their life as used engines. Availability — and to some extent cost — is a strong factor here. On any given day there are literally tens of thousands of used engines for sale in the United States. Most wrecking yards are linked by a teletype or similar communication system which facilitates finding what you want in the area.

If you go to a wrecking yard to buy an engine, you know it is used before you ever set eyes on it. What you don't know is how used it is. If at all possible try to locate the engine you want in a wrecked car, as opposed to one that has already been removed from a car. The condition of the car can tell you a lot about the condition of the engine. If the vehicle appears to have been well maintained up to the time it was wrecked, chances are the engine was reasonably maintained also. When you really get interested in a specific engine, press the wrecking yard to fire it up for you. They can do this — even if the engine has already been removed from the vehicle. For a fee most yards can deliver an engine to you if you live in the immediate area. Most yards will also guarantee the major parts of an engine to be mechnaically sound for 60 to 90 days. If this is the case, get it in writing on the bill of sale. Unfortunately, the word and handshake of western man is a shadow of its former self.

If the engine you lay out three to five hundred scoots for turns out to be a sick turkey, what do you have to do to prove this to the wrecking yard and what do they do about it? Do they give you your money back or do they give you another engine? Do a lot of shopping for engines, prices, conditions and guarantees. For the most part there are more engines around then there are buyers so you don't have to be pressured into buying.

In addition to what we've already pointed out, there are several common sense areas to be followed when buying a used engine. Stay away from any engine damaged by fire. Check all engines carefully for physical damage. If the oil pan was opened up during a long skidding wreck, the engine may have run out of oil while still running while at the same time ingesting dirt and gravel. In the long run this could be a very expensive used engine. Is the oil filter intact? It should be. A damaged crank pulley or dampner could mean the crank received a severe blow which in turn could cause expensive internal damages. The same thing goes for the tailshaft of the transmission. If a vehicle has been whacked severely from the rear, the axle can move forward and shove the driveshaft far enough into the transmission to ruin the transmission, so watch for this when buying an engine and transmission together. Which brings us to the next point:

Buy Engine and Transmission Together

The reasons should be fairly obvious. If you are a klutz, you will not have to unbolt one component from the other. Secondly, you have the unspoken assurance that the factory meant the two components to go together. Having been there before I can tell you it does not make much (if any) sense to jump around from one wrecking yard to another in order to get the "long deal" on a transmission here and an engine there. If you figure your time is worth anything at all, then buy a complete engine and transmission together and get on with building the rod.

ENGINE CHOICE

We are not history buffs in the sense we stay up nights reading about the Russian revolution or how Davey

In normally aspirated form, the little Buick is smaller than any V8, and the power output is more than adequate for a 2,500 pound car.

Crockett got his at the Alamo. But, as all historians are quick to tell us, it is important to know where we have been in order to have a decent grasp on where we are going. Theory of history in hand, let us proceed directly to hot rod history.

The first hot rods in this country were Model T Fords. They were plentiful, cheap and the aftermarket suppliers jumped on the bandwagon in quick order to supply everything from two speed rear axles to tilting "fat man steering wheels." Then came the ever popular Model A which really caught on big in the hot rodding world when Ford came forth with the V8 in 1932. By the late thirties, the tip was to have a V8 in a Model A — an A-V8. By the early fifties, engine swapping was a major satellite interest of hot rodding. With the advent of the overhead valve V8 from every major domestic car builder by the mid-fifties, engine swapping was at the forefront of hot rodding in this country. In 1942, the hot tip was to put a '40 Ford V8 into a Model A. By 1952, the hot tip was to put a Studebaker, Olds or Cadillac overhead valve V8 into a '40 Ford.

By 1962, if you wanted reasonable cost and reliable high horsepower that could almost be serviced at the corner drug store, your rod would be equipped with a small block Chevy. The same held true for 1972 and will almost hold true for 1982 — almost, but not quite.

These are changing times, and street rodders will have to change with them. As this is being written (mid '82) we are kissin' close to a buck-and-a-half a gallon for regular gasoline, more and more stations are not selling premium, and those that are, sell a far inferior grade than found at the pumps ten years ago. Cars are getting lighter, engines are getting smaller. The factories are faced with something called CAFE — for Corporate Average Fuel Economy. The net result of this will be all sorts of subtle and not so subtle marketing chicanery to convince you to buy a V6 or four banger instead of the optional V8.

For those of us brought up on V8 high performance engines, none of this will be very easy. We can rant, rave, pound our fists on the kitchen table and carry on at length about being sold down the river by the "goviment", "Detroit-City," "do-gooders for clean air", and the "oil producting states." The fact remains that by 1985, the engines rolling out of Detroit will be V6s and four bangers. Turbochargers and diesels will be common-place.

By 1985, we predict you'll still be able to buy a V8 — new or in a wrecking yard. The small block Chevy, Ford and Chrysler V8 will then be regarded as the big block Chevy, Ford and Chrysler are today. Thus, if you are contemplating building a repro rod in the early to mid eighties, then we suggest you give some serious thought to powering the mechanism with V6 or in-line four cylinder power.

Naturally, the latter suggestion might not go down easily; so you are certainly within your rights to rant, rave

and pound on the aforementioned kitchen table.

So were does this leave you for an engine for your repro rod?

Tradition dictates that an American hot rod be powered by an American-built engine. Obviously this narrows the choice considerably. At this writing the most obvious pretender to the throne now occupied by the small block Chevy is the even fire Buick V6. This engine was and is in the right place at the right time — beginning with the great fuel crisis of late 1973. The Buick V6 is manufactured by the Buick Motor Division of General Motors and sold to all other GM passenger car divisions. It is available in a number of displacements and either normally aspirated or turbocharged. Current production figures are 5,500 complete units per day which means that by the beginning of the 1985 model year about nine million of the little devils will have been built.

Although the Buick V6 may be comparatively rare (to the Chevy V8), production figures such as those just cited have a way of evening the balance in a hurry — especially in light of the fact that small block V8 production will be tapering off. Thus, we will stick our neck out on the chopping block even further and say that by 1985, *the engine to have in a street rod (repro rod) will be a Buick V6.*

TRANSMISSIONS

Elsewhere in this book, we suggest that you buy the engine and transmission as a unit if at all possible. This way you get a transmission that the factory intended to go with the engine — and which in fact, does bolt up. Although this may seem to be pretty simple stuff, it can prove to be a nightmare to the first time rod builder. For instance, will all small block Ford V8s bolt to all C4 and C6 transmissions? Will all GM Turbo 350 or 400 automatics bolt to any late small block Chevy? If not, why not? Can

Here's the backside of a Buick V6 installation with a Turbo 350 clearing the firewall and floor with room to spare.

the problem be solved with an adapter? How much is the adapter and where do you get one? Never mind what Herkimer at the wrecking yard says; does it fit?

Because repro rods are light and are generally not used to tow heavy trailers up mountains, there is no need to beef up a stock transmission with trick modulators, shift kits and so-called street/strip clutch packs. Like a modern domestic engine, an automatic transmission is good for 100,000 miles of use unless it has been punished or abused. Drop the pan, change the oil and the filter and enjoy street rodding instead of trying to "out trick" yourself.

It is imperative that a cooler be used with an automatic transmission. Our thoughts on this are under the heading of RADIATOR (later in this book).

By now, you may have noticed all of our remarks about transmissions concern automatics and not a word about standard transmissions. The reason for this is simple. For the average repro rod builder (and the professional), a standard transmission is a pain for a number of reasons. Not necessarily in order of pain it causes:

1. The shifter never comes up through the floor where you want it. This can be endured, or it can be remedied by hours of fabricating brackets and linkage.

2. The guy who really wants to buy your rod at top dollar does not know how to drive a stick shift (don't laugh) or your wife, girl friend or both think your car is cute but they keep hammering on your skull that you were dumb to build it with a stick shift trans. This gets old in a hurry.

3. Clutch linkage from start to finish is time consuming beyond belief for most rod builders. If you do build it correctly and there are absolutely no problems under any conditions, no one will appreciate it. If you have some problems (like the clutch not releasing fully under acceleration or high pedal effort) everyone will just think you're a dummy for not having built the car with an automatic.

4. As a final word of discouragement relative to stick shift transmissions in repro rods, consider that the firewall and cowl area of a repro rod are constructed of fiberglass and thus will have to be extensively braced if a swing clutch and brake pedal assembly is to be used. The world if full of misery and work; why make more of it?

If you are reading this chapter to find a hopeful word or two about using engine/trans adapters in order to install a '65 Olds engine in a '40 Ford coupe, then you are reading the wrong book. There is a considerable difference between building a repro rod and a "nostalgia rod." The book deals with the former and not at all with the latter. The reason should be self evident — there is a considerable difference in the level of skill needed between the two areas of interest.

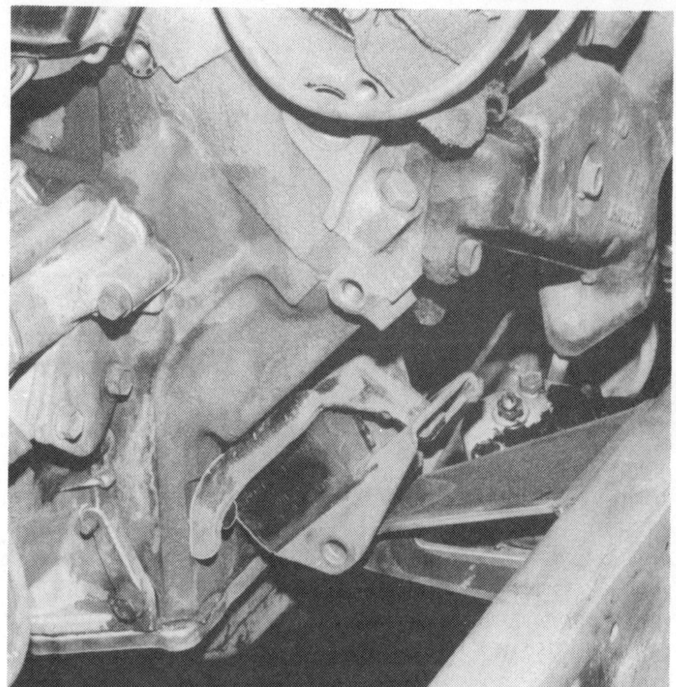

Stock rubber mounts should always be used in a street rod to minimize noise and vibration. Manufacturers invest hundreds of thousands of dollars in engineering and testing production mounts — why not take advantage of their expertise.

MOUNTS

Automotive engineers spend a lot of time using sophisticated instruments trying to develop an engine mount which does all the things a mount has to do — hold the engine in place while limiting movement and dampening the amount of noise and vibration the engine transmits into the car. The placement of the mount on the engine is studied, the design of the mount is analyzed and hundreds of prototype mounts are constructed and tested. The durometer (hardness) of the rubber in the mount is altered along with the design as the testing and narrowing down proceeds. With this as a background, we've never understood why rod builders and engine swappers preferred to completely re-engineer existing engine mounts.

One of the worst approaches to take in building engine mounts for anything which will be driven on the street is to mount the engine solidly to the frame. This transfers all vibration, noise and torque into the frame, which in turn breaks all sorts of things while at the same time feeds vibration and noise into the car, and to a certain extent, the bodies of the occupants.

A quantum leap up from the solid mount approach is the '32-'48 Ford rubber biscuit "hot rod" mount. These mounts have been around since Henry installed a V8 in the first '32. They are still very popular with street rodders and they seem to be chosen because: a) they are "hot rod" and b) they look neat (because beauty really is in the eye of the beholder, we reserve comment).

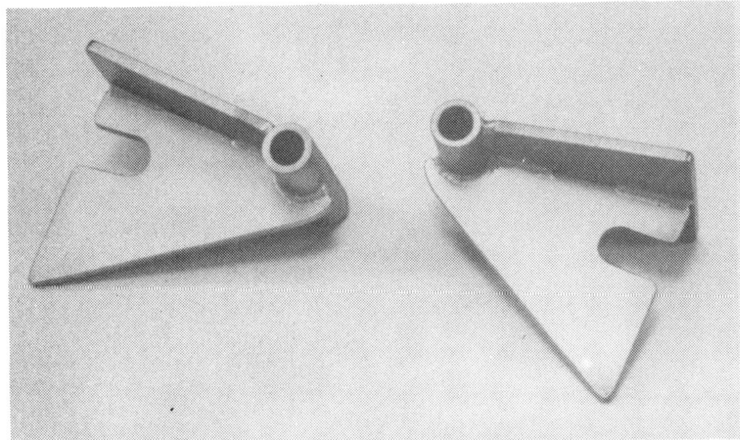

Challenger, T.C.I. and some other manufacturers offer these pre-welded mounts for simplifying the installation of most any engine into a street rod. Notch allows the passage of plumbing and wiring.

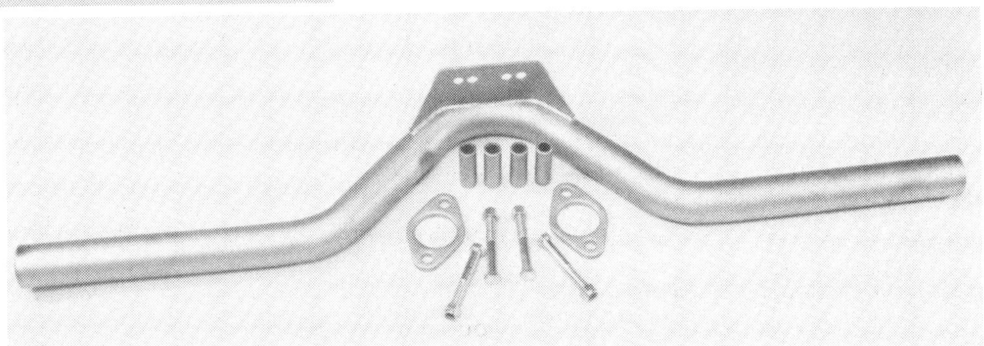

Bolt-in crossmembers are also readily available from a variety of manufacturers. By all means, the car should be fabricated so the transmission can be pulled without having to remove the engine.

We like stock mounts because of the engineering behind them, the fact they are readily available and virtually trouble free for the life of a car — and we're talking in terms of 150,000 miles in a production car. Contemporary mounting practice dictates that an engine/transmission assembly be mounted at one point on each side of the block and at one point at the rear of the transmission. This is the practice with all contemporary GM, Ford, Chrysler and AMC passenger cars — presumably they are not all wrong.

EXHAUST SYSTEM

Depending on the repro rod being built and your skills as a fabricator, an exhaust system can either be duck soup or a disaster — rarely is it anywhere in between. To begin with you should know the entire system will have to be custom built in every respect from behind the headers or exhaust manifold. Serious thought need not be given to the exhaust system until the entire chassis has been constructed and all running rear is in place. It is imperative that all brake lines, crossmembers, ladder bars, rear shocks and springs and bumper brackets be in place before ever thinking about constructing an exhaust system.

Weird, multi-muffler, solid-mounted, chrome-plated exhaust systems have a way of not returning value to the rod owner comensurate with the cost. Gaining in popularity with rod builders are practical exhaust systems

which are moderate in cost and will live nearly forever. An increasing number of rod builders are retaining stock exhaust manifolds for reasons of cost, quietness and the fact that most headers on most street rods do not give any mileage or performance benefits due to the way the car is normally driven.

One method of improving the appearance of a stock exhaust manifold and provide an extremely durable finish is have it porcelainized. This is a ceramic coating which

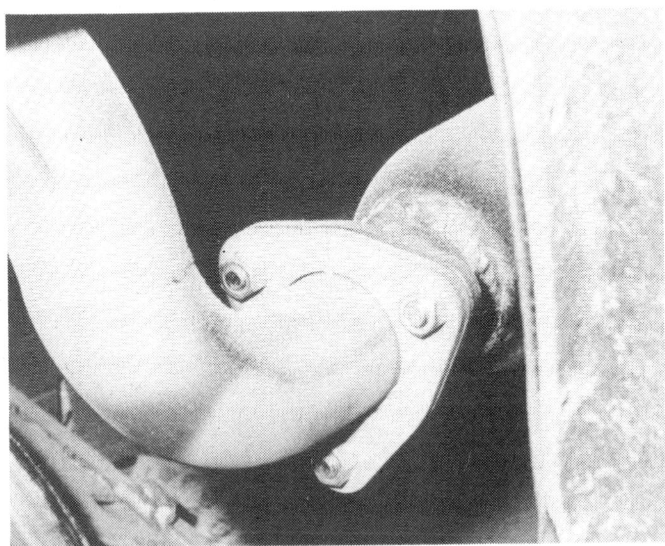

One very practical way to constructing an exhaust system is to build it in sections and then attach the sections with header flanges.

Porcelainizing the entire exhaust system is a good investment for a street rod. It will not discolor like chrome, nor will it chip or flake.

Space is a premium commodity under a street rod as well as in it. Here is the Magoo approach to routing the exhaust around a ladderbar assembly.

adheres to the cast iron. To prepare a manifold for this treatment, grind off all sharp edges and non-essential numbers and casting marks, glass bead the parts and start searching the Yellow Pages in a large metropolitian area for someone who can do the work. The rest of the exhaust system can also be porcelainized if the system is put together in sections with flanges at the ends of each section. Muffler shops have a wide variety of flanges — the three bolt variety provides maximum clamping. Don't get fancy with the hardware mounting the exhaust system to the chassis. Use standard rubber muffler and exhaust pipe hangers to reduce the amount of vibration fed into the chassis.

For all of the reasons we suggest that might lead you to retain the stock cast iron exhaust manifolds, the same holds true preferring a single exhaust system over dual exhaust. One of the problems encountered when building an exhaust system for a repro rod equipped with a rear 4-bar kit, coil-over shocks, etc., is that you simply run out of room under the car to hang battery box, master cylinder, power brake booster and a dual exhaust system. Oval-shaped modern-day passenger car mufflers (even the shorty Turbo Corvair) are very difficult to fit under a modern rod. Think in terms of 2-inch OD exhaust and tail pipe and something on the order of a Mitchell glass-pack muffler 30-inches long bearing part number PNBHMB30. This particular muffler has been certified to California sound standards.

Chrome headers and chrome cast iron exhaust manifolds turn an uneven blue due to the heat. If you have no objection to this and want chrome here, by all means chrome it — we just didn't want you to be surprised when your high dollar chrome hardware begins to take on the appearance of faded blue bloomers about ten minutes after you light the fire.

An exhaust system for a repro rod should be given as much thought as the rest of the car. It should be practical, durable and attractive in that order.

DRIVESHAFTS

Due to the countless variations of engines, transmissions, rear axles and placement of all of these items within various chassis, you can count on having a custom driveshaft built for your repro rod. This is relatively simple and

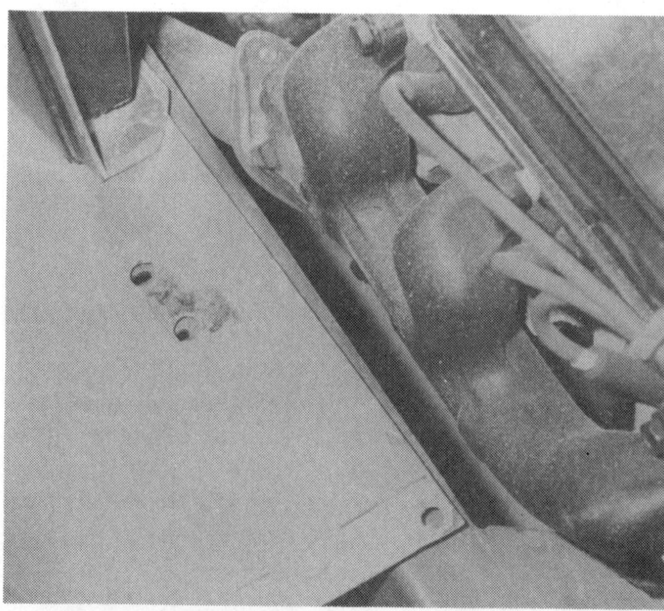

When stock engine mounts are used, make sure there is at least ½-inch of clearance between all exhaust components and any part of the chassis or running gear. This exhaust pipe later caused problems due to insufficient clearance.

inexpensive in any part of the country, but here's some background on the subject which should make it even easier.

Give some thought to the driveshaft when you buy the engine and transmission by asking for the driveshaft that was mated to the transmission. Do the same thing when purchasing the rear axle. If all goes according to plan you will now have two driveshafts. Let's say you're using a Buick V6 and Turbo 350 in the car and a Mustang rear axle. One driveshaft will be compatible with the GM running gear, the other with Ford running gear. Simply hang on to both driveshafts until the engine, trans and rear axle are installed in the chassis. At this point simply shove the driveshaft compatible with the transmission into the back of the transmission until it won't go any further and pull it back exactly ¾-inch. Now the center-to-center distance between the front and rear U-joint crosses needs to be measured with a steel tape. Make sure the tape is tight, and make sure the vehicle weight is resting on all four wheels. Note the measurement and take both driveshafts to a machine shop and they take over from there. If the machine shop or driveshaft shop (there are shops which do nothing but driveshaft work) do a lot of automotive work, you may be able to simply tell them what transmission and rear axle you have in addition to the center to center dimensions and have them come up with the correct driveshaft.

Driveshafts are everyday jobs for machine shops all over the country; they are used in cars, trucks, back hoes, corn binders, combines, streetsweepers, ditch diggers and drilling rigs just to name a few of the applications.

If for some reason you need any part of a driveshaft assembly to aid the machinist in the construction of a special driveshaft, you need only to check with a large auto parts firm. TRW, Spicer and Borg Warner all have catalogs devoted to driveshaft hardware. From these catalogs you can obtain everything needed to construct a driveshaft — including the tubing.

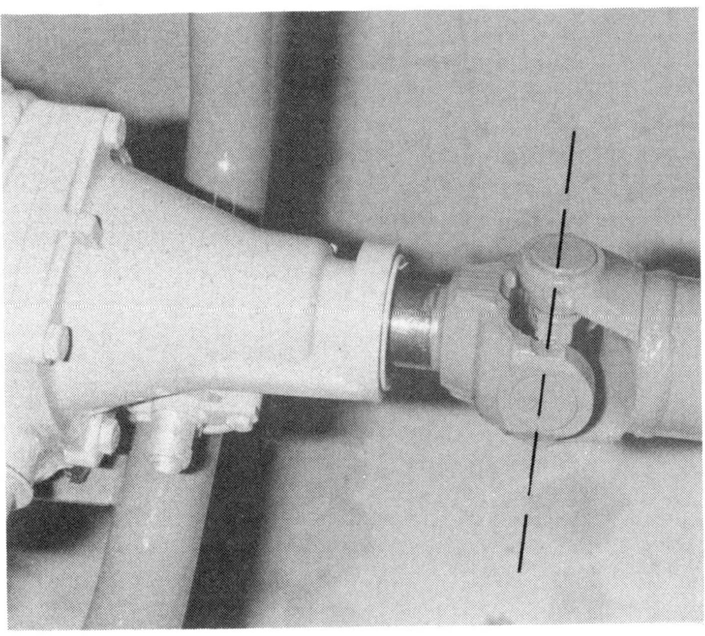

To measure a driveshaft for length shove the yoke into the transmission until it bottoms out, then pull it out one inch. The measurement you want is from the center of the front yoke to the center of the rear yoke.

Cooling

RADIATOR

Like other pieces of reproduction hardware, shop for quality — not price. How much difference can there be between two repro radiators for a Model A? Well, how about the lower mounting holes being off ¾-inch, or a soldered joint around the upper neck leaking, or there being no provision for the support rods? Sure, all of this can be fixed, but why? The plan in building a repro rod is to take it out of the box, have it fit, bolt it on and get on with the project.

If you will be using an automatic transmission, consider ordering a repro radiator with a trans cooler built into the lower tank. This is a clean, sanitary way of solving the trans cooler problem. We are not impressed by the advertising copy for all of the aftermarket trans coolers. GMC, Ford, Chrysler and AMC have used the lower tank to house their trans coolers for years with no particular problems. If for some reason you cannot get or do not have a radiator with the trans cooler incorporated in the lower tank and the car will be running an automatic transmission, make sure that the aftermarket trans cooler is mounted away from the exhaust system and that it is exposed to a flow of fresh air when the car is moving.

You should know that people like Walker Radiator have radiators for early cars in a variety of core thickness for optimum cooling. In the case of Walker, they also have radiators with an air conditioning condenser built in. This is the most sanitary of all approaches when building an air conditioned early car.

An overheating street rod is no more fun than an overheating production car or truck. Putting a large engine behind an early vintage radiator will usually lead to problems when you least want them.

Adapting electric fans from small foreign cars to a street rod almost never works out. The problem is lack of air moving ability on the part of the fan. This may be fine on the highway, but keep your eyes glued to the temperature gauge while caught in traffic in Tucson in August!

The most popular method of cooling the engine of a repro rod currently is to install a thermostatically controlled electric fan manufactured specifically for street rods.

You should also know that a core too thick for the job can cause overheating simply because it is more difficult to pass air through a 4-inch core than it is a 3-inch core. The bottom line is to stick with reputable, well established manufacturers who know more about solving overheating problems than your buddy at the local radiator shop who plans to cut down a radiator out of a Peterbilt because "he knows" it will keep your rod engine cool.

FANS

More important to cooling than the type of fan used is the proximity of the fan to the radiator. If you don't run a shroud (and few rods do) the leading edge of the fan should be within 1-inch of the back of the radiator. I like ¾-inch even better. Most auto parts stores have spacers in ½-inch increments which allow the fan to be moved forward or back on the water pump hub. A flex-fan isn't quite as effecent as a seven blade viscous clutch fan. The fixed seven blade fan offers the most cooling potential.

Relative to the centerline of the crankshaft, most early engines — like Model A and flathead V8 Fords — had cooling fans placed fairly high compared to the fans on modern overhead valve engines. This creates a problem when a late engine is mated with an early car. The fan is positioned to pull air through the lowermost portion of the radiator but not the top — thus the cooling capability of the radiator is seriously impared. Over the years, brackets, pulleys and kits have been devised to raise the fan and move the air through the majority of the radiator for maximum cooling.

The problem can be avoided and the solution simplified by the use of an electric, thermostatically controlled fan which attaches directly to the radiator core. The electric motor can be actuated manually or thermostatically from engine coolant heat. This makes a lot of sense because it means the fan is running only when it is needed. Many current production cars have adopted this method of cooling fan operation. The system is simple and troublefree and relatively inexpensive.

An electric fan assembly from a late model car can sometimes be adapted to work in a rod, but most of the time this is more time consuming and trouble than it is worth. There are aftermarket kits available which you can take right out of the box and install.

THERMOSTATS

Changing thermostats is rarely (if ever) a cure-all for an overheating problem, but some time spent in this area may be beneficial if overheating problems are encountered. For years the most common thermostat was the 180-degree model. Now most of the engines feature 195-degree thermostats; 205-degree thermostats are also available. This forces the engine to operate at higher temperatures and this in turn is beneficial in controlling emissions. Unfortunately this narrows the safety gap between normal temperature and overheating. The increased temperature also makes an engine more prone to detonate. None of the problems, events or consequences associated with overheating amuse me in the slightest, which is why most every vehicle in my driveway or garage is outfitted with a 180-degree thermostat. For turbocharged engines you might want to consider something in the area of 160 degrees.

Wheels and Tires

The tire and wheel combination used on a rod is an extremely important part of achieving the overall visual impact you want the car to project. The tires and wheels can account for as much as one half the total height of a car and up to one third of the length. They can be used to control or alter the width as well as control the atitude or rake of the car. When on the car, and when planning what to put on the car, the package of wheels and tires cannot be separated, but for discussion purposes they must be.

TIRES

From a practical standpoint the only tire to be considered for a modern street rod is a radial with street-type tread. This eliminates the once popular sprint car tires and the current crop of very agressive off-road truck tires. A modern radial lends its construction characteristics to a smooth ride with a beam axle front suspension and live axle rear suspension. The radials are very long wearing and puncture resistant. Their cornering capabilities are far superior to that of a bias or belted/bias tire.

Tire buying — especially performance tire buying — has become extremely complicated in the last ten years. New size designations have appeared; there are now several basic size coding systems with each being slightly different from the others. About the only thing the various systems seem to have in common is tire width and wheel diameter. Other elements are added to each system to indicate those characteristics not common or basic to tire construction.

Tire width is designated in at least three ways:

Decimal — This is the oldest designation still around and it is now used on bias construction tires only. Examples of this are 6.00, 6.70 and 8.55 × 15.

Metric — Now we're back down to radial construction only and the dimensions are being expressed in millimeters such as 165, 225 and so on.

Letter — This designation is used on both bias and radial construction tires. A, C, H, etc., are examples of this.

Outside of some rare exceptions, wheel diameter is still expressed in inches — 13, 14 and 15-inch will get you through the basic course of street rodding.

Aspect ratio refers to the percentage of sidewall height to width such as 50 series, 60 series, 70 series and so on. The aspect ratio number will always be present with the letter designation of width and it is sometimes present with the metric designation of width. Aspect ratio affects tire diameter because sidewall height is concerned. As the aspect ratio gets smaller the tire gets wider, but it also gets smaller in diameter. A G70 will be wider, but smaller in diamter (shorter) than a G78.

A comparison of the most common tire aspect ratios. The lower the number, the wider and lower the tire. Tread width is proportioned to the tire width.

A tire of radial ply construction always exhibits an R on the sidewall regardless of how width is expressed — BR70-13 and P215/50R-13 are two examples.

A performance rating is sometimes present in metric radials. Currently there are two ratings — S for standard speed usage (up to 112 mph) and H for high performance usage (up to 130 mph).

On metric radials you'll find the prefix "P." This stands for pressure rating which means 3 psi more inflation pressure than standard recommendation will very closely align the metric radial with the nearest size letter radial. For example, in the case of P195/70R-13, an additional 3 psi can be added to achieve the same static loaded radius of a BR70-13.

Although we've included some engineering data on some specific tires from B.F. Goodrich that falls within the needs of rod builders — along with a basic cross reference table — there are several general comments to make concerning tires on street rods. To begin with, all of the tire engineering data in the world will not give you the final answer about the tires you want on your rod due to what you desire in the way of overall appearance. While your car is under construction you should go out of your way to attend rod runs to search for cars similar to yours in search of ideas. When you find one that pleases you in the tire and wheel department, write down exactly what the tire and wheel combination is. Find the owner and see what his thoughts are on the subject. Do the rear tires rub the body during cornering? Do the front tires scrape the underside of the fender when a bump or dip is encountered? Would the owner buy and install the exact combination again? What is the rim width, what is the offset? Were there problems getting the wheels? Did these wheels create other problems such as clearance with the brake caliper?

Tires tend to look larger when standing alone than when mounted and installed on a car. You should know there can be up to an inch difference in width or height of a modern radial between two brands of tires even though the indicated size is the same. This is why performance tire people will refer to one brand of tire as fat and another brand as skinny. Appearance is a characteristic of the brand. Another characteristic is how flat or curved the tread is compared to another tire.

Here are two prime examples of cars tucking all of the front tire and wheel combination under the fender.

Perhaps more than any other area of visual concern, the tire and wheel combination on a rod is responsible for your first and most lasting impression of a car. This "big 'n little" combination with a very wide, flat tread in the rear is close enough to what a sprint car runs to lend an air of competition to a street-driven car.

In sharp contrast, a full fendered phaeton with the rolling hardware much less aggressive could pass as a restored car in some circles. The impact of the tire and wheel has been lessened which in turn heightens the impact of the body lines.

RADIAL T/A™ (70 SERIES)
ENGINEERING DATA (269 Stock Series)

Tire Size	Stock Number	BFG Recommended Rim Width Range	Inflated Dimensions (Inches) @ 24 PSI*					Rev./ Mile @ 45 MPH	Ctr. Skid Depth 32nds	Load Range B	
			Design Rim Width	Static Loaded Radius	Section Width	Overall Diameter	Tread Width			24 PSI Load (lb)	Max. (lb) Load @ 32 PSI
175/70R13	269-301	4.5-6.5	5.0	10.16	6.94	22.66	5.16	897	11	890	980
185/70R13	269-302	4.5-7.0	5.0	10.70	7.39	23.49	5.50	876	11	990	1090
185/70R14	269-303	4.5-7.0	5.0	11.14	7.40	24.37	5.39	851	11	1045	1155
195/70R14	269-304	5.0-7.0	5.5	11.26	7.71	24.89	5.67	837	12	1155	1280

Tire Size	Stock Number	BFG Recommended Rim Width Range	Inflated Dimensions (Inches) @ 26 PSI (180 kPa)*					Rev./ Mile @ 45 MPH	Ctr. Skid Depth 32nds	Standard Load	
			Design Rim Width	Static Loaded Radius	Section Width	Overall Diameter	Tread Width			180 kPa (26 PSI) Load (lb)	Max. (lb) Load @ 240 kPa (35 PSI)
P195/70R13	269-631	5.0-7.0	5.5	10.50	7.70	23.61	5.78	871	12	1080	1246
P205/70R13	269-623	5.0-7.5	5.5	10.84	7.99	24.34	6.07	850	12	1179	1356
P205/70R14	269-626	5.0-7.5	5.5	11.20	7.99	25.18	6.00	825	12	1235	1433
P215/70R14	269-632	5.5-8.0	6.0	11.42	8.39	25.67	6.26	813	12	1345	1554
P225/70R14	269-633	5.5-8.5	6.0	11.65	8.71	26.19	6.54	797	12	1455	1675
P235/70R14	269-634	6.0-8.5	6.5	11.87	9.17	26.69	6.82	783	12	1565	1808
P215/70R15	269-639	5.5-8.0	6.0	11.94	8.50	26.88	6.26	778	12	1411	1620
P225/70R15	269-636	5.5-8.5	6.0	12.10	8.71	27.21	6.51	769	12	1521	1753
P235/70R15	269-637	6.0-8.5	6.5	12.39	9.25	27.85	6.84	751	12	1642	1896
P255/70R15	269-638	6.0-9.5	7.0	12.76	9.97	28.67	7.32	728	12	1885	2183

RADIAL SPORT TRUCK T/A™
ENGINEERING DATA (428 Stock Series)

Tire Size	Stock Number	BFG Recommended Rim Width Range	Load Range	Dimensions at Max. Inflation (Inches)*					Rev./ Mile @ 45 MPH	Center Skid Depth 32nds	Max. Load and Inflation
				Measuring Rim Width	Static Loaded Radius	Section Width	Overall Diameter	Tread Width			
10R15LT	428-378	7.0-9.0	C	8.0	14.15	10.90	30.80	7.90	680	16	2230 @ 50 PSI
10R16.5LT	428-385	8.25	C	8.75	14.15	10.75	30.80	7.90	680	16	2330 @ 50 PSI
12R15LT	428-539	8.0-10.0	C	10.00	14.70	12.90	32.40	9.40	641	16	2850 @ 50 PSI
12R16.5LT	428-520	8.25-9.75	C	9.75	14.70	12.40	32.40	9.40	641	16	2370 @ 35 PSI

Obviously this engineering data does not cover the entire line of BF Goodrich tires. The data on the two series of tires shown will cover 90% of all street rod applications when the desire is for "littles on the front and bigs on the back." Obviously similar data is available from the manufacturer of all other tires.

Although the look is now found nationwide, this '32 three-window exhibits the classic "California look." The rake of the car, the tire wheel combination, the completely filled rear wheelwell all contribute to the look.

Fenderless cars have somewhat more freedom in a tire/wheel combination in that fender clearance problems do not exist. On the other hand the tires and wheels are much more responsible for how the car appears. These wires, dropped large diameter headlights and exposed drag link steering all work together to give this car an "early rod" look. We suspect it is very new.

Here is a very practical, subtle combination tucked under the fender of a '32 sedan. This is nit-picking but notice how the wheel/tire is biased ever so slightly to the rear of the wheelwell . . .

. . . whereas this five-window coupe has the axle biased ever so slightly to the front of the wheelwell.

WHEELS

Spend a lot of time looking at other rods and rod magazines to determine exactly what you want. Discussing wheels to be used on a rod is sort of like getting into a dialog about the physical characteristics and merits of someone's wife — with her husband. Do a lot of thinking before getting in over your head.

Do you want the wheels to be visual accents on the car, or do you want them to blend in and be a subtle part of the overall view? Some wheels have a way of instantly dating a car — is that what you want? If you have a fendered car do you want to tuck all of the tire and wheel under the fender? Some guys do, some don't. Obviously the tire and wheel combination can contribute to the overall attitude of the car. Rims of 14 inches in the front and 15 inches in the rear are very popular, but is that what you want?

If you are having custom wheels built you will have to know what bolt pattern you want, rim width, rim diameter and off-set. Take a look at the two drawings and you'll be able to determine off-set wanted — dependant on where you want the tire in the fender.

If you want an aftermarket chrome, mag or aluminum wheel, talk the situation over with other rodders in your area. Where did they get their wheels? Were they satisfied with the service? Would the dealer mount up a tire and wheel for you and put it on your car to see if you were happy with the appearance?

Take your time. Wheels are difficult to wear out and used ones can be difficult to sell — especially if you're the guy trying to sell them so you can raise the money to buy what you really want.

Generally speaking a tire of radial construction has a much more rounded tread area which gives a more subtle look generally than . . .

. . . a bias or bias belted tire. This is an early MoPar truck rear wheel with a lot of positive off-set which moves quite a bit of the tire out of the fender. Like the rest of the car, this is a matter of personal taste.

MEASURING BOLT PATTERNS
— THE EASY WAY —

Here's a neat little trick for determining the bolt pattern of a five lug wheel. Measure from the back of one stud to the center of another one opposite the first. This will give you the bolt pattern measurement every time. If it is a four or six bolt pattern, measure from center to center of opposite studs (or holes).

This is how "offset" or "backside setting" is measured. It is the distance between the backside of the wheel center to the outside surface of the rim.

THE NO-SPARE SYNDROME

There is precious little room in a street rod, and the chances are good that you won't have room in yours for a fully mounted spare after throwing in several changes of clothes, a small cooler and maybe a sleeping bag. If you pack a cover for your rod, suddenly you've got a full trunk. If you're the nervous type you may be somewhat concerned about not having room for a spare tire. Forget it. In the first place, the chances of you having a flat on the road with a good set of tires today is rare indeed. If this is not sufficient comfort to you and a 3000-mile trip in the rod is being planned, consider this: Pack the smallest scissor or hydraulic jack that will safely lift and hold your rod. Pack a lug wrench or socket and breaker bar of the correct size to remove and replace either front or rear wheel and tire. Pack a valve stem removal tool, a spark plug engine air compressor hose and the necessary wrenches and fittings along with a medium size bottle of tire sealant.

Despite advertising claims to the contrary, we can't get excited about running squirt-in tire sealant as a matter of routine due to the possibility of difficult or impossible imbalance problems.

If and ever you do have a flat, place the flat tire in the position of having the valve stem at the bottom of the tire pointing straight up. Jack up the car until the offended wheel is clear off the ground. Remove the valve stem, and squirt in the contents of the tire sealant container. Rotate the tire by hand for as long as the instructions on the tire sealant container indicate. Don't cheat here. Replace the valve stem and inflate the tire with the spark plug engine air compressor hose.

Brakes

Like all of the rest of the hardware going into a street rod, everything available in the way of brakes has been strapped to a street rod at one time or another in the last thirty years. At one time it was the hot tip just to have hydraulic brakes on a Model A. Then came the era of adapting very large drum brakes to all kinds of hot rods. At one time T-buckets with rear brakes only were being built and so on ad nauseum.. Fortunately it appears the choices in the '80's are mostly good and mostly revolving around full size passenger car front disc brakes and rear drum brakes.

Probably the most popular choice of brake hardware today revolves around the hardware made by Super Bell Axle Company. At the top of the list when it comes to popularity and stopping power is their Mustang disc brake kit which mates the vented factory rotors and single piston factory calipers to '37-'48 Ford spindles or the Super Bell spindles. The hubs that come with the kit are available in 4½, 4¾, 5 and 5½-inch diameter bolt circles which allow tremendous selection of wheels.

In addition to the Mustang brake kit, Super Bell also has a line of caliper brackets, rotors and hubs which allow the use of Hurst/Airheart dual calipers on a Ford spindle, VW Type IV calipers on a Ford spindle, and VW Type III calipers on a Ford spindle in addition the same named calipers on '49 to '54 Chevy spindles. All of this comes without machine work or "maybe we can get it to work if we just heat and bend this a little."

There is some hardware around which will allow converting late GM and Ford rear axles to disc brakes, but as far as we know all of this hardware requires welding or

Super Bell Axle Co. makes this "Low Buck" brake kit which allows the use of most '70 thru '77 intermediate GM calipers and rotors. The kit is engineered for cars weighing more than 3,000 lbs.

machine work — and in most cases both. Disc brakes are neat for street rod front brakes because the front brakes do most of the work in any application, but primarily disc brakes are neat for the front of a street rod because they are more readily available in wrecking yards than drum brakes.

The third brake kit offering from Super Bell is what they term "Super Stoppers." They supply all shown here with the rod builder being given the choice of selecting two different VW calipers or two different Hurst/Airheart calipers.

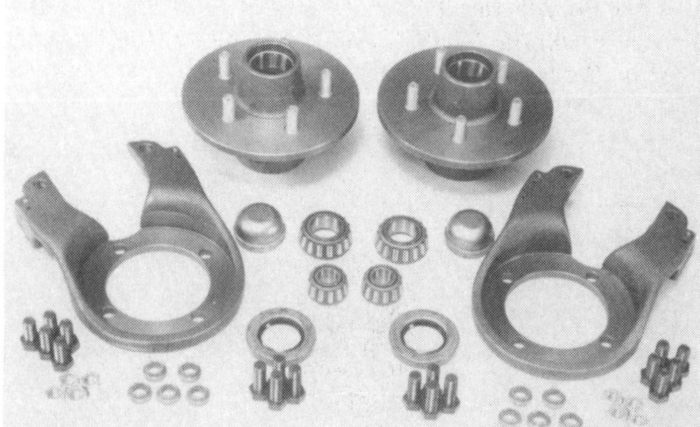

(Above and Below) Probably the most popular brake kit on the market today for repro rod construction is the Mustang brake kit manufactured by Super Bell. They supply hubs, bearings and caliper brackets. The rod builder supplies Mustang calipers and rotors. When used with Mustang drum brakes in the rear and a power booster, braking is superb.

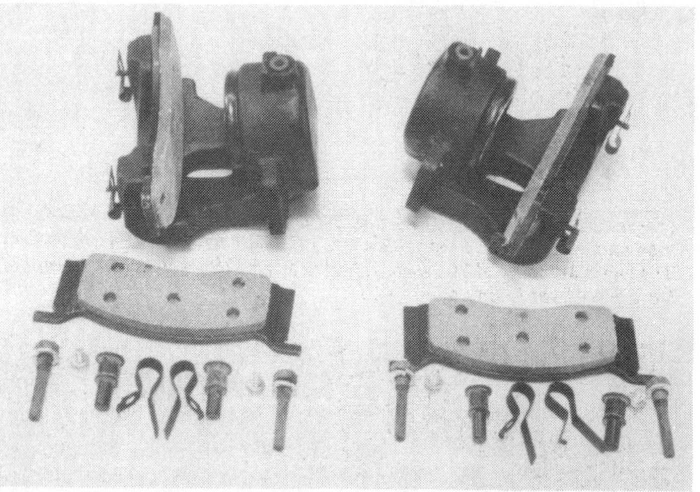

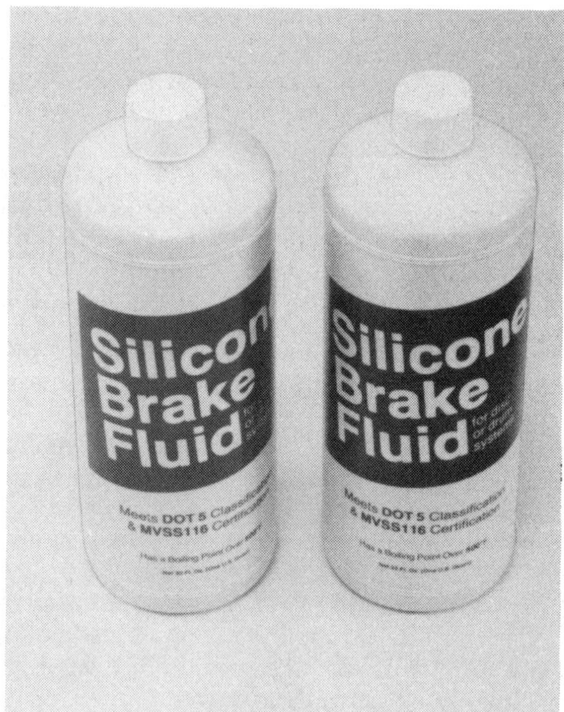

Although silicone brake fluid is a superior product, its usage in a street rod is more than justified because it will not remove the paint, which regular fluid will.

(Left) Super Bell offers front brake line kits for VW, Mustang and Hurst/Airheart calipers. This is probably the most sanitary setup available.

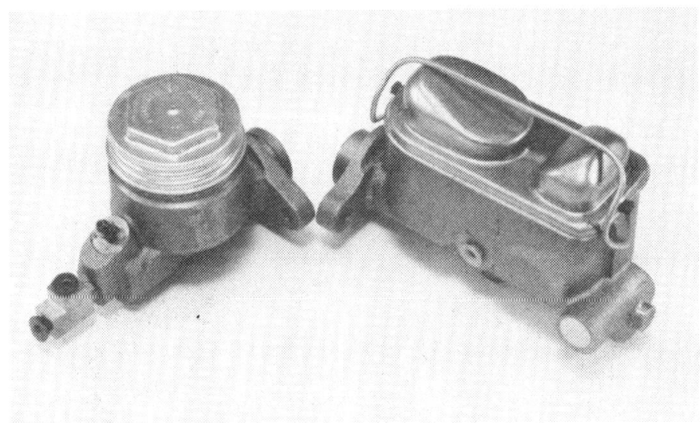

At left is the EIS E39626 master cylinder. At right is the EIS E71248, which is a dual cylinder for front discs and rear drum brakes.

PEDAL RATIO

Pedal ratio is very important to the amount of leg effort needed to get a car stopped. Pedal ratio is the measurement of mechanical leverage built into the brake pedal which determines how much pedal pressure is needed to create a certain amount of pressure at the master cylinder. For example, a 100-poiund force applied to a pedal with a 5:1 ratio will result in a 500-pound force input at the master cylinder. Generally speaking, the greater the pedal ratio, the less effort required to produce the same effect.

MASTER CYLINDERS

One master cylinder which works quite well with the Mustang front disc and drum rear brakes is the ⅞-inch bore EIS unit bearing the part number E39626. Remove the residual pressure valve from inside the cylinder of a single type cylinder for drums if discs are being used. Cylinder E39626 was intended to be used for all an all-drum system, but works quite well for discs if the check valve is removed.

When selecting a master cylinder, you should know that a ⅝-inch bore is easier (softer) to operate than a 1-inch. This is because a larger bore moves more fluid — but at less pressure — when the pedal effort remains the same. A 1-inch bore master cylinder should be power boosted when used with discs.

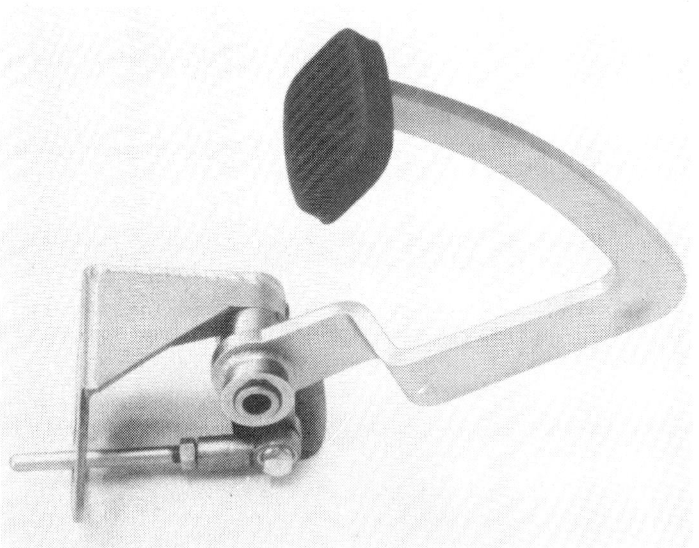

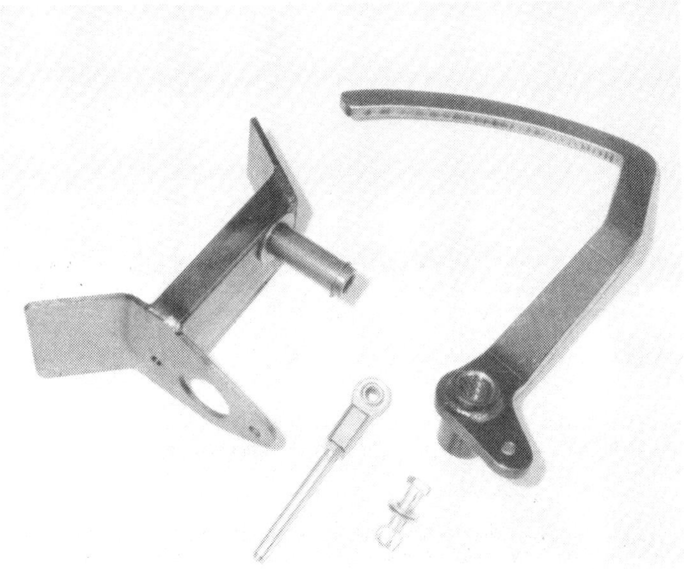

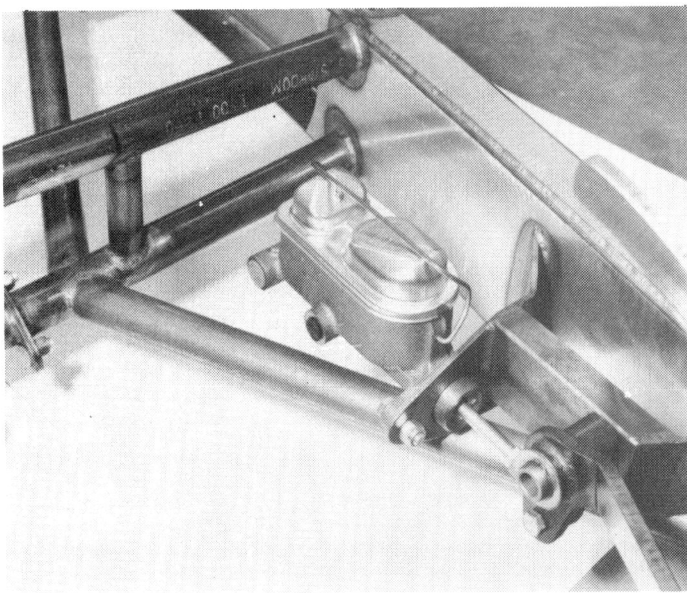

(Above and Below) Brake pedal assemblies and master cylinder mountings should be left to those fabricators who have done all of this before. These are samples of TCI work. They're top-of-the-line in every respect. Note the offset in one of the pedal arms to clear the steering column.

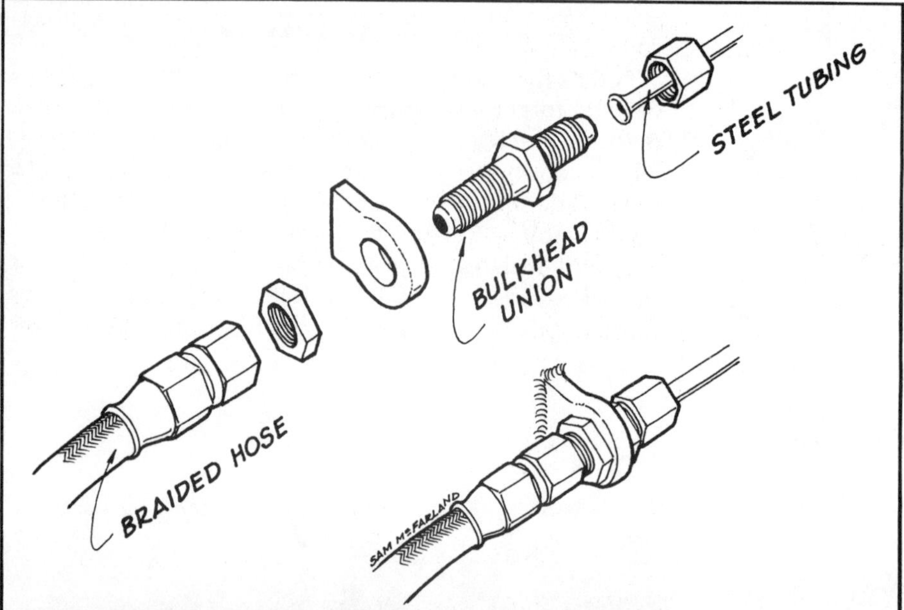

(Left) One really neat way of making the transition from rigid steel line (never copper!) is to weld a tab to the frame or crossmember and then use an aircraft-type bulkhead fitting to join rigid and flex lines.

(Below) Unless you plan on doing some substantial bracing on the inside of the firewall, mounting the master cylinder here is a very bad deal.

This is a good example of why a professional chassis builder should be handed as much of the fabrication at once. The guy who built this was really surprised when the dual reservoir master cylinder hit the crossmember and could not be used. Later it was discovered the pedal arm interfered with the steering column and guess who mounted the steering box.

Even on a fenderless car don't get so hung up on the appearance of a brake assembly package that you sacrifice safety for appearance. These are all good examples of plenty of brake. If you are making up your own assembly, make sure the caliper is mounted so the bleed valve is at the very uppermost position of the caliper so the caliper can be bled.

At rod meets you attend while your car is going together, take a good look at pedal positions and assemblies. Depending on your car and your size, something like this might be just what you are looking for.

From a practical standpoint, a 5:1 pedal ratio should be considered the minimum. There is a formula for determining pedal ratio:

$$\frac{A \times B}{C}$$

A = Required line pressure
B = Area of master cylinder piston
C = Desired pedal effort

Example:
A = 1200 pounds (average line pressure for disc brakes)
B = .6016 (area of ⅝-inch master cylinder)
C = 100 pounds (average pedal pressure applied)

$$\frac{1200 \times .6016}{100} = 7.21{:}1 \text{ pedal ratio}$$

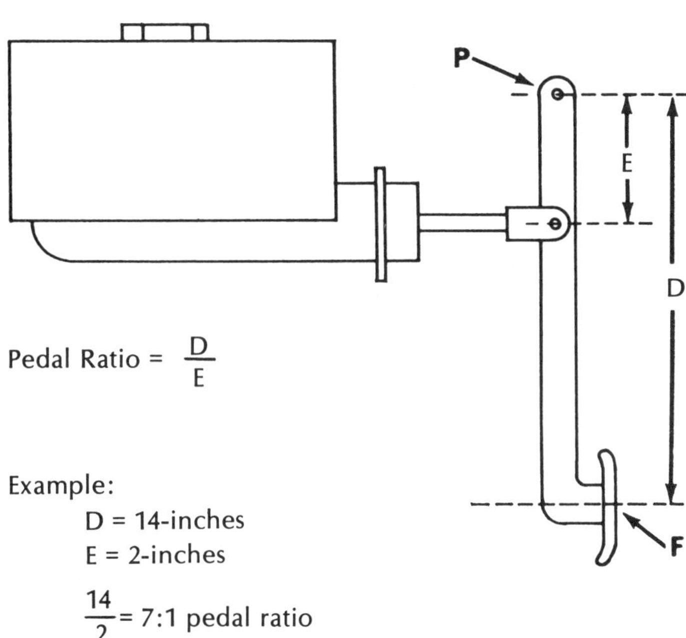

Pedal Ratio = $\dfrac{D}{E}$

Example:
D = 14-inches
E = 2-inches

$$\frac{14}{2} = 7{:}1 \text{ pedal ratio}$$

POWER BOOSTERS

Before telling yourself you don't need a power brake booster on your repro rod, there are several things to consider. In the first place, all cars sold today with disc brakes and an automatic transmission installed at the factory come equipped with power brakes from the factory as standard equipment. That should tell you something. Compared to drum brakes, disc brakes

(Above and Below) This is the Midland Ross remote power brake booster as mounted in two different cars. If the car is being built with disc brakes, this item should almost be considered mandatory.

require an extreme amount of pedal effort. You've probably heard or read this before and simply tucked the information back into the corners of your mind. Believe it. Disc brakes and automatic transmissions do not mix without power brakes. One power brake booster which is gaining in popularity with street rodders is the Midland Ross unit bearing the part number of SC 3500BC. The unit can be mounted to the inside of the frame or to a crossmember completely hidden from view. This is a remote unit plumbed into the brake line immediately after the master cylinder before the lines split off toward the front and rear brakes. This unit cannot be used with a dual type master cylinder — this is a cylinder having a separate line for front and rear brakes.

BRAKE PROPORTIONING

Most repro rods using disc front and drum rear brakes don't need a proportioning valve to achieve the correct bias between front and rear. The exception to this is a

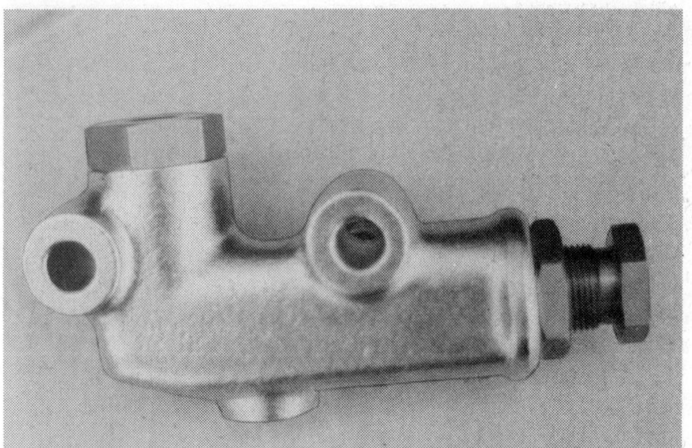

This is the Super Bell brake proportioning valve which regulates line pressure.

A modern emergency brake assembly is a minor problem in a repro rod simply due to lack of available space. Here is a GM H-car (Vega) assembly. You can run out of foot room in a hurry when both the shifter and emergency brake are located on the floor.

lightweight, early pickup which, due to lack of weight on the rear axle, needs some definite help in preventing the rear brakes from locking up far too early. Kelsey-Hayes used to build an adjustable proportioning valve which was very popular with rodders but this little piece of cast iron is mighty rare today. If you cannot locate one, consider the manually adjustable piece from Super Bell Axle Company.

Not so incidentally, when you have the car running and are gradually "debugging" it and taking care of such cut-and-dry items as adjusting brakes, work up to speed slowly, trying out the brakes in increasing increments of 10 mph. Pick an open area with little or no traffic and plenty of room to recover from skidding or spinning. As soon as the front brakes lock, all control over steering is lost. A wheel must be revolving on the pavement before it can be steered.

EMERGENCY BRAKE

Not only is it illegal, but it is also very dumb not to have an emergency brake on a repro rod. You may go a lifetime and never use it, but the one time you do need it, I can assure you that you will really need it. Keep the emergency brake as simple as the rest of the car. In production cars they are simple, very straightforward mechanical devices employing leverage, a ratchet and a cable to keep the rear brake shoes pressed against the drums. All sorts of problems arise when fitting an emergency brake to a disc brake rear axle — one more reason to keep it simple.

Modern emergency brakes are mounted adjacent to the driver's side kick panel and are actuated by foot, or they are mounted on or near the driveshaft hump in the floorboard and are actuated by hand. Pinto, Vega and Corvette emergency brake units are popular floor mounted units for repro rods. They are readily available and relatively easy to mount. That's the good news. The bad news is that with this hardware, the emergency brake

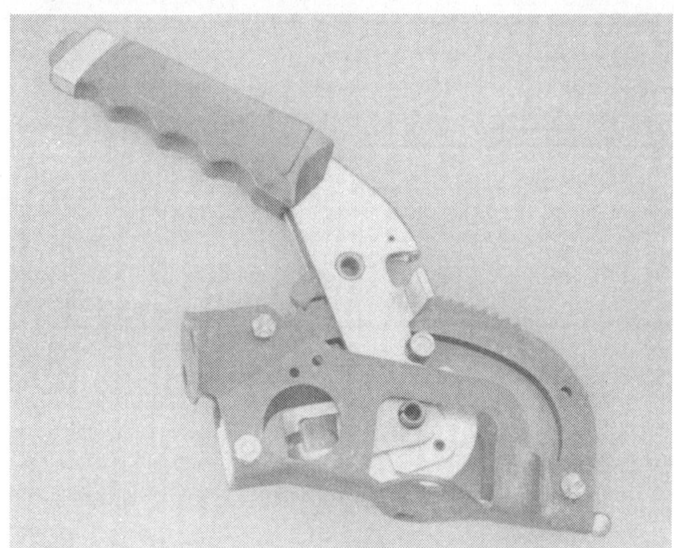

This is the emergency brake assembly from a late Corvette. This is more compact than the Vega item but obviously more difficult to mount.

wants to live where the transmission, seat and floorboard are already placed, so depending on the powertrain and the body being used you can run out of room for a floor mounted emergency brake in a hurry.

There are at least two readily available, low cost foot-actuated emergency brake assemblies relatively easy to mount and use in a street rod. Both are available new or in wrecking yards. The first one bears the GM part number of 344626 and uses GM cable assembly 338250. These parts are used in '73-'79 GM X-bodied cars — Apollo, Nova, Omega, Ventura and Skylark. The second emergency brake unit which often finds a home in a repro rod is hardware from a '73 Ford pickup. The pedal assembly unit has a Ford part number of D3TZ-2780-A. The part number of the cable assembly is C5AZ-2853-A. Brackets to hold either foot-actuated assemblies will have to be fabricated. If you want a dual rear wheel emergency brake, visit your local wrecking yard for the appropriate yoke hardware and mounting brackets.

Single wheel emergency brakes are a lot simpler and are used all the time by professional rod builders. If you need to clamp one cable to another during the fabricating of the system, a U-type electrical clamp can be used. These are readily available at industrial electrical supply firms. If you need an end sweged onto a shortened cable, visit a local aircraft repair firm or a boat yard.

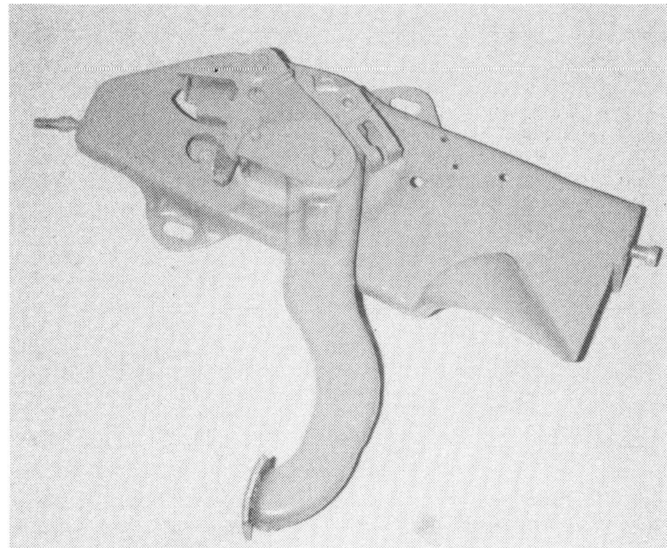

(Left and Above) This is the '73-'79 GM X-bodied car emergency brake assembly mentioned in the text. With a small amount of fabricating this unit is about as easy to mount as any.

(Right) This shows the relationship of steering column to brake pedal on a relatively straight-forward Model A-based repro rod. Pedal pivot should be equipped with a bronze bushing and grease fitting.

The Repro Body

Reproduction fiberglass early Ford bodies are available from a number of sources around the country. In addition to the half-dozen or so firms that actually manufacture the bodies, they are sold in rod shops and by various mail order outfits. In addition to the bodies, there are reproduction fenders, running boards, grille shells, seat risers, seat assemblies, hoods, splash aprons, rumble seat assemblies and all of the hardware (door hinges, latches, striker plates, etc.) that it takes to put together the body for a repro rod with all new pieces.

There are two distinct levels of quality — one we will term "competition," the other we will label "quality." This is not to say that the competition level of body is not quality. It is simply of a different quality than that you want on a repro rod. A competition body is considerably less in terms of cost than a body suitable for use on a repro rod. The competition body is lighter and has little or no internal structure. There is no floor, the doors are a molded in as part of the body and there is little or no wood or steel glassed to the inside of the body.

Manufacturers such as Wescott, Gibbons and Poliform build bodies which are heavy, stiff and feature pre-hung doors, complete floors and hinged deck lids or rumble seats. These pieces are considerably more expensive initially than a competition body, but in the end they are a far better buy for the amateur rod builder.

The problems associated with trying to upgrade a competition body are endless. Wood or steel or both is glassed into place on the inside of the doors and on the inside of the body around the door opening. Then the door panel is cut out of the body with a saber saw. Hinges, latches and striker plates then get mounted. There are problems in getting the doors to open and close properly. When the hapless builder figures it is "good enough," the body is then set on the frame and attempts are made to build a floor of plywood by bonding it to the body with fiberglass. At this point it is discovered the lower edge of the body is considerably narrower than the frame it is to fit on. Before the floor is cut, the body is jacked outward to meet the frame. Of course when this is done, the doors spring open and cannot be made to latch. Six beers later and a lot of cussing, it is reasoned that the body "must" fit the frame and the doors can always be rehung later. With the body fitting the frame (well, almost), the doors are forgotten and the splash aprons are set in place. They don't fit very well either, and it appears that jacking the body out at only one point instead of three or four before the floor was glassed in had something to do with it.

Get the picture? The average rod builder who goes this route in constructing a repro rod is well on his way to joining our all-too-common "95% Club." Friends are called in to analyze the situation. They all agree it is pretty

(Left) This is Dick Williams of Poliform with just some of the reproduction body pieces he manufactures in addition to complete old Ford bodies.

(Below) One of the neat things about working with repro hadware is that you may not want every piece to be an exact reproduction. These are two versions of a Poliform '32 dash which will fit into the Poliform Model A body.

All of the molds to make a high quality repro body represent a considerable investment in terms of the many man-hours which must be invested to make each panel perfect before it can be used as a mold. If the manufacturer is to offer choices on a particular item such as dash and recessed firewall, the investment increases.

(Right) A mold must be perfectly slick and waxed if a part is to be released. Note fiberglass overspray on the sawhorse.

A chopper gun is used to spray on a combination of fiberglass strands and laminating resin. Then it must be carefully shoved into every edge of the mold. This is the outside of a Poliform Model A door. When this piece is cured, it will be trimmed and bonded to a fiberglass inner door panel.

Heavy ribbing is used on the backside of the mold to reduce warpage.

Here Dick Williams and Jim Kirby of Challenger being prying the mold off a roadster body. If the body doesn't pull cleanly, the mold and the body will be damaged.

terrible and that the body is junk. For a week or so some friends of our rod builder drop by after work to help a little or just check on progress being made. Soon interest fades. The friends believe that building a repro rod is a very cruel, practical joke played by the manufacturer on the unsuspecting gullible rodder, and any thoughts they might have harbored about building a repro rod are gone forever. Perhaps our rod builder is the persistant type and decides to seek professional help. On the surface, two avenues are open — a good body shop or a shop specializing in fiberglass boat repair. But at this point the water gets even muddier — the project of making a fiberglass car body "right" in either shop is an oddity. Rip-off artists in either business will take the rodder to the financial cleaners and still not solve all of the problems. A competent shop will keep track of their time for billing purposes, do the best they can and apologize in the end when the job is terminated.

In most cases the project never gets this far along. The rod building stops and body is sold for practically nothing (to another amateur who knows he can fix everything), it is cut up and thrown out in the dead of night (in embarrassment) or it all simply sits in the garage gathering dust.

None of this is a very pretty picture of a guy who just started out to build a neat old hot rod for fun, is it?

It does not have to be this way. Consider the case of the street quality repro body. At first blush they are very expensive. With only a little experience on the part of the rodder, they become solid bargains. Doors are hung, latch hardware is in place, the rumble seat or trunk lid is cut and hinged, the floor is in place, along with several options of dash and firewall. All manufacturers of street quality bodies previously mentioned have their own methods of making sure their bodies fit repro frames.

The message is clear — to build a repro rod, buy the very best you can the first time around. If you are still concerned about price at this point, consider the racer's adage — when you buy the best, you only cry once.

If this message of shopping quality instead of price is beginning to sink in — just keep thinking that way and never look back. For purposes of making a point let's assume you decide to build a full fendered '29 roadster with a street quality body. You have resigned yourself to paying "full pop" for the body and all the attaching panels; but in a final, last ditch effort at economy you write one check for the body, one for the front fenders, one for the rear fenders and one for the splash aprons. You have flipped through all of the catalogs, totaled up all of the costs and figured out how you can save $116 on the total package. Dumb, really dumb!

Let's say the body fits the frame fine. The doors open

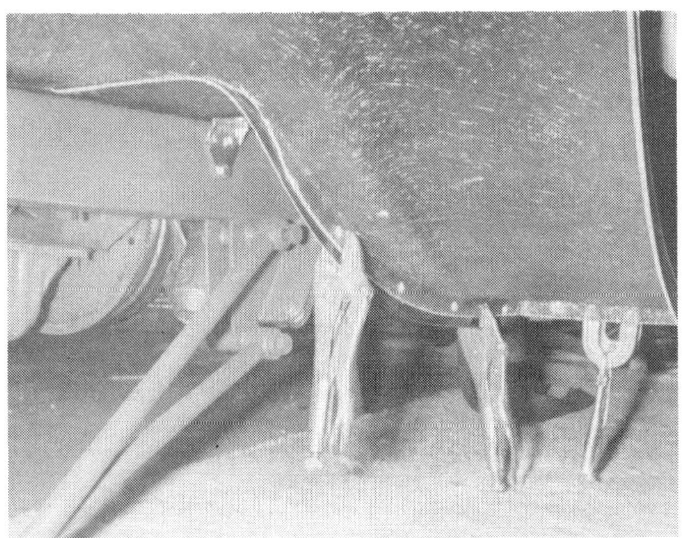

One of the problems you'll encounter when building a repro rod is that everything seems to need to be done at the same time and nothing can be done until something else is done. In this case vise grips are being used to clamp a front fender to a splash apron and running board to check for fit.

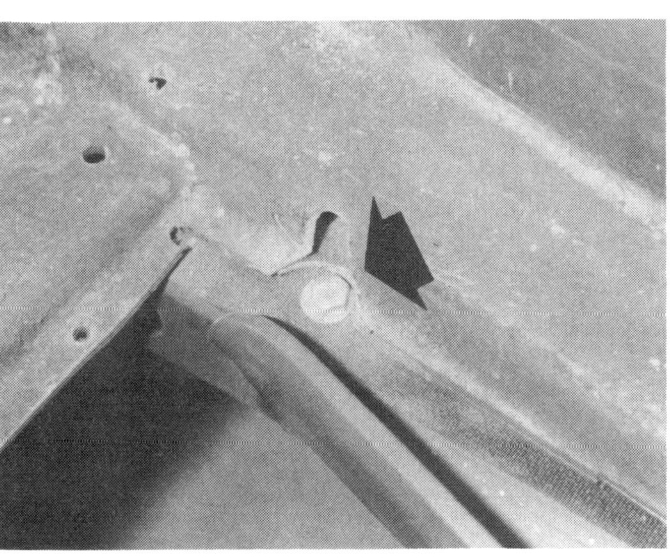

No matter what brand repro body you buy, it will have to be modified — count on it. Here's a case where the body had to be notched to provide clearance for the bolt which holds the rear bumper bracket to the frame.

Here's another common area of body surgery. Hot rod type coil-over rear shock assemblies most often interfere with the floor area to the rear of the crossmember. As work progresses this notching can be sealed off with a fabricated metal or fiberglass piece bonded to the floor.

and close quite decently. However, there is an ugly gap between the front of the rear fender and the body. When you drill the necessary holes and pull the fenders into place, the trailing edge of the fenders points outboard. Up front, a similar situation exists: no matter what you do, the leading edge of the fenders points inward, and after the kind of dough you've spent you're not about to put up with a pigeon-toed repro rod. You're mad and you want to complain. You also want the problem solved. Who do you call — the manufacturer of the body, the front fenders, the rear fenders, or the running boards which tie it all together?

Let's say you call the body manufacturer and complain about the fit. In short order they will ascertain you have an X body, with Y rear fenders and Z front fenders and that you did all of this to save $116. Now you are complaining to them about gaposis in the fitting department. With this sort of complaint, just how far do you expect to get?

The solution to (or avoidance of) the problem should be painfully obvious — buy all repro body components from the same manufacturer! If you have complaints or questions you have only one phone call to make. Obviously, the manufacturer has his work cut out for him if you can honestly tell him that all of the components were built by him and that they do not fit.

SPLASH APRONS

A special word is in order about fiberglass splash aprons. Although fiberglass splash aprons are readily available, most professional rod builders and savvy amateurs stay away from them. Splash aprons fit between

The amount of wood and/or steel used to brace a repro body varies among the top manufacturers. Don't be afraid to augment what the manufacturer has done whether it be for purposes of structural improvement, ease of construction (seat construction) or aesthetics — such as upholstery.

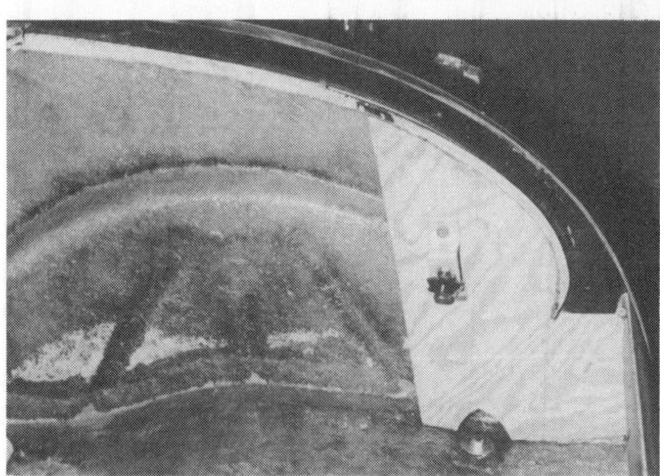

Obviously some wood can serve a multiplicity of purposes — in this case a hinge location point for a rumble seat, a stiffening member for the rear of the car and as an attachment point for an upholstered panel.

If you intend to spend any effort in augmenting the structural integrity of a repro body — start with the area around the door. Stiffness behind the hinges is most important for trouble-free door closure. The bottom line is to insure the hinge attachment points can't move around regardless of whether the door is latched or unlatched.

Adding structural integrity to a glass repro body is pretty easy. Wood or steel can be laminated to the body with glass cloth or mat using laminating resin — all supplies are readily available a boat shops or large paint stores. Simply follow the simple directions.

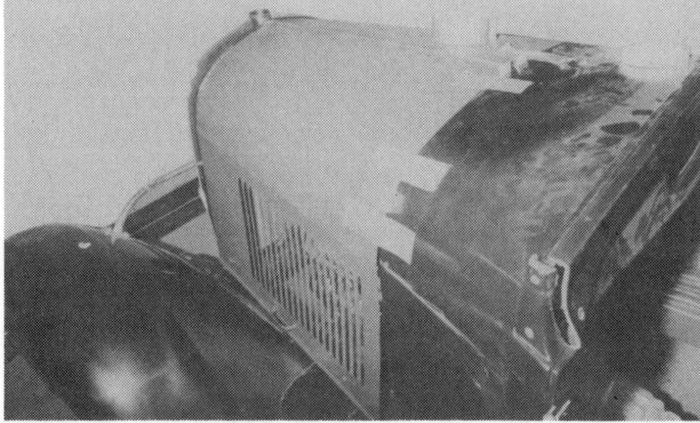

Every panel on a repro rod should be assembled and aligned before any panel is readied for paint. This shot was taken fairly early in the process of getting everything to fit.

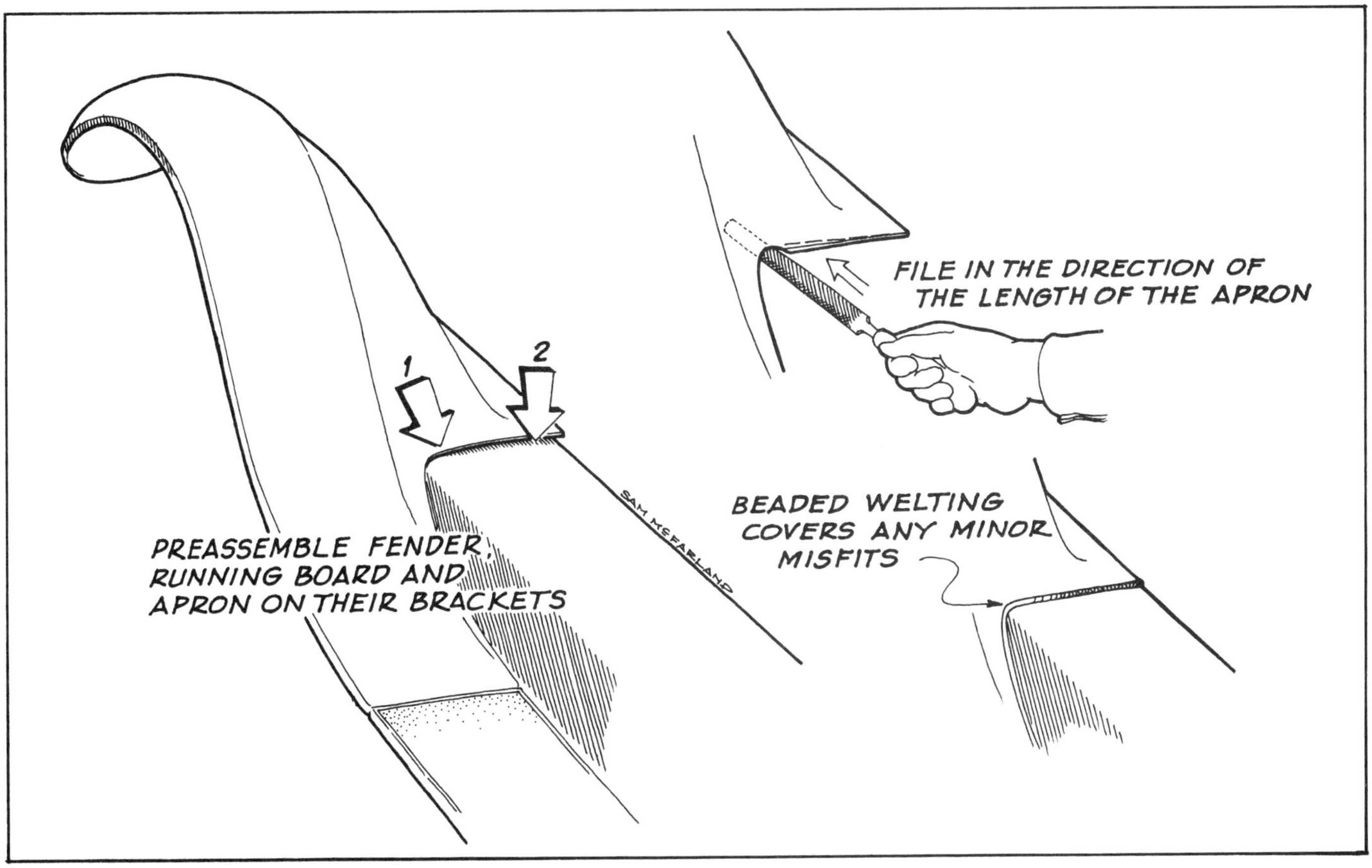

The area between the fender and splash apron often needs work for a perfect fit. In this case some filing and grinding is needed in area one to tighten up area two. Grind and file is the plan. A saber saw is just too much temptation to hurry the job up — and possibly ruin it.

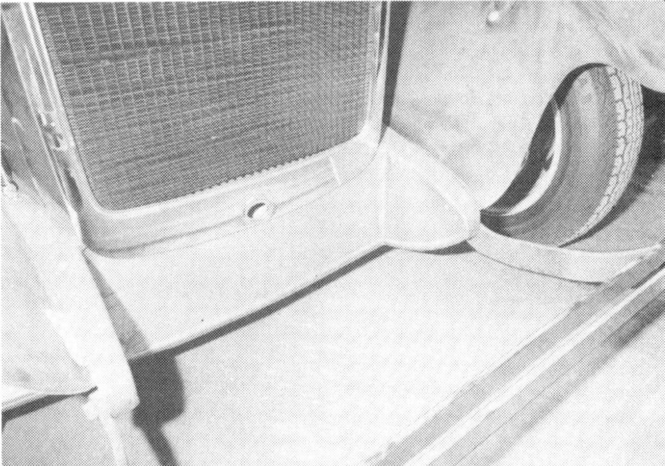

If the car is to have bumpers, install whatever you plan to use, measure it to make sure it is level with the world. With this accomplished, remove the bumper and brackets until final assembly. Even in a large shop these bumpers can be "shin grabbers."

On a fendered Model A with reproduction hood this is how the hood latch gets anchored.

the frame and the body. Fiberglass aprons are three to five times as thick as steel repro items. This "non-stock thickness" makes it almost impossible to get the hood to fit without raising the radiator and grille shell. The fitment problems continue — use steel repro splash aprons and avoid the problems.

DON'T GET FANCY

If you are a first or even second time car builder you'd best think twice before getting fancy and making alternations to the existing lines of the car. For instance, using a '32 Ford grille shell on a Model A has been popular for years. It appears to be a bolt-on. Before trying this you might like to know that the Model A radiator is too wide to accept the '32 grille shell and that the '32 radiator needs to be shortened to allow the hood to slope slightly downward from rear to front to make the silouette of the car "read correctly." For the experienced craftsman and rod builder this one seemingly simple modification can turn to a nightmare costing hundreds of dollars as a custom radiator is built and then a custom hood is fabricated. Rest assured that for the inexperienced builder, any step out of the ordinary will most likely lead to a disaster.

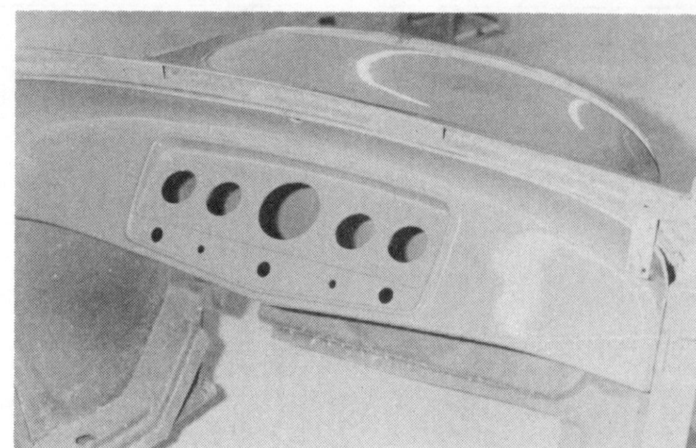

See what a difference a paint job can make! For best appearance this should now be rubbed out before any gauges, switches or other panels get in the way. This is the way the entire body should be treated.

(Right) The preliminary build up of the car should not stop with getting the body to fit correctly on the frame and having all panels in correct relationship to each other; everything that it takes to make a complete car should be taken care of at this stage except wiring and upholstery — and you should be giving some thought to those items at this stage.

(Below) A radiator apron on a repro rod can be a troublesome item to fit on a repro rod. Like splash aprons, some rod builders recommend using only metal repro items here.

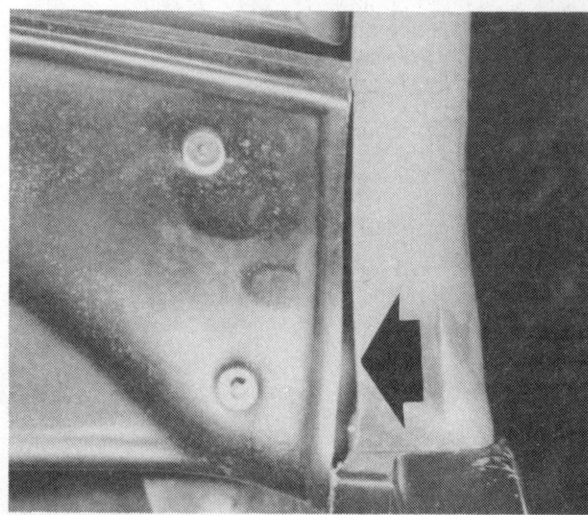

Gaps such as this around the dash and windshield posts are not uncommon. Depending on how much this bothers you it can be left there, filled with fiberglass, the post can be built with metal and ground to the correct shape, or welting can be used to the hide the gap.

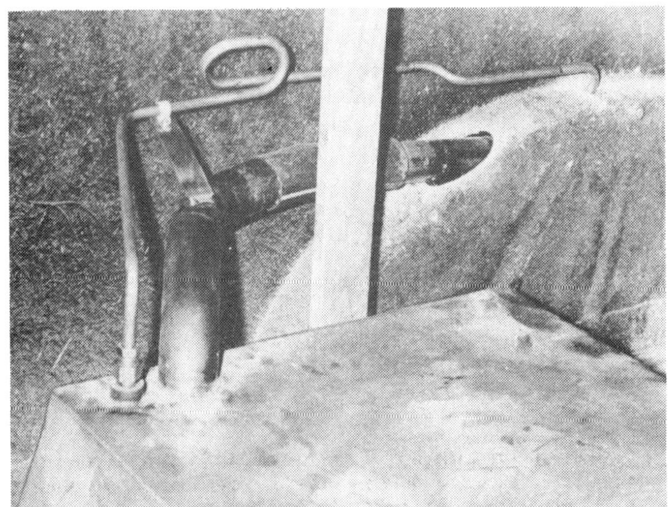

Here's one way of filling and venting a fabricated gas tank in a roadster. When designing and executing a major part of the car such as this, never lose sight of how the entire system can be removed, serviced and replaced. If this particular shot seems confusing, maybe we can help by telling you the gas tank filler is in the top of the right rear fender.

Here's another view of that gas tank which rides immediately behind the seat back. Baffles are most important; the larger the tank the more important.

Don't overlook the obvious; make the attachment of all hardware as simple and straightforward as possible. You may not have to unbolt anything major from the car for ten years — but when the day comes that you do, you may not want to spend a week doing it.

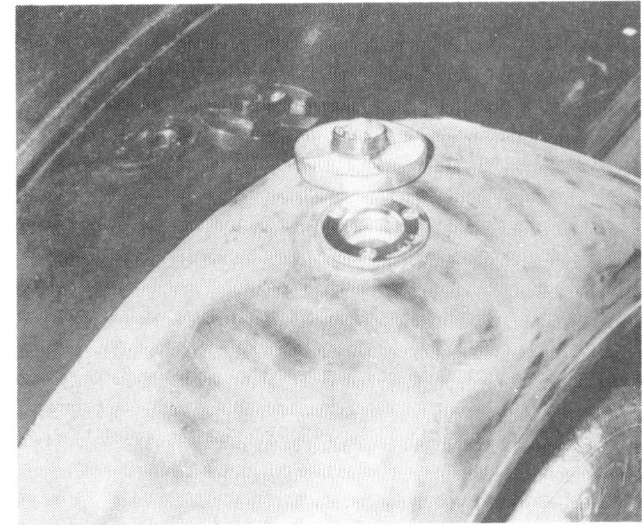

Here is a view of the rear fender location for the gas fill. A boat fuel fill bung has been grafted to a Model A rumble seat step pad. Due to the ninety degree turn which is needed immediately under the fender, filling is very slow.

(Left) Regardless of whether you adapt a wrecking yard gas tank or have one fabricated you should thoroughly clean it with lacquer thinner or acetone, let it dry and then liberally coat the inside of the tank with a commercially available fuel tank slushing compound. RV stores and airplane supply stores are good sources for this.

Wiring and Instruments

WIRING

Although it is easy to wire a car with the accompanying diagrams, it is boring and extremely time consuming. Professional rod builders like Magoo and Jerry Kugel use Ron Francis' ready-made wiring looms for hot rods because they work and they save immense amounts of time. Every connection is marked and color coded; all of the harness assemblies include headlight, dimmer and ignition switches, ignition resistor, horn relay, and turn signal flasher. You can order harnesses which incorporate wiring and hardware for every conceivable rod accessory — air conditioning, radio, back-up lights, power seats, etc. The instructions are easy to follow and make sense.

If you use the Ron Francis' wiring harness, you should keep the instructions and make them part of your permanent notebook of facts on the car.

Even if you do not opt to use a complete wiring harness from this supplier, one of his catalogs is invaluable in finding what you need in the way of hard-to-find wiring hardware. See the suppliers list for more details.

Use tie-wraps to keep the wires in a bundle, or you can string the wiring job out forever by lacing all of the wires together with fishing line. If you use the Ron Francis modular fuse/terminal panel, about the only decision you'll have to make is where to put it. There are two logical choices — on the floor under the seat, or under the dash on the firewall. The firewall is less accessible than under the seat, but then how much routine or emergency work do you plan on doing to the wiring? Leaving the area under the seat unencumbered allows the storage of all sorts of handy items such as fire extinguisher, jack, tools, NSRA Fellow Pages and enough rags to keep all of this rubbish from rattling.

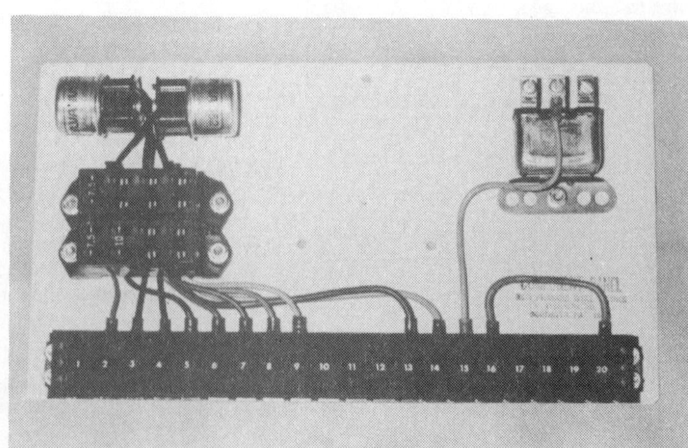

This is the heart of the new Ron Francis wiring module which is so simple and straightforward that using this plus the appropriate wiring loom from the same supplier saves many hours in wiring a car.

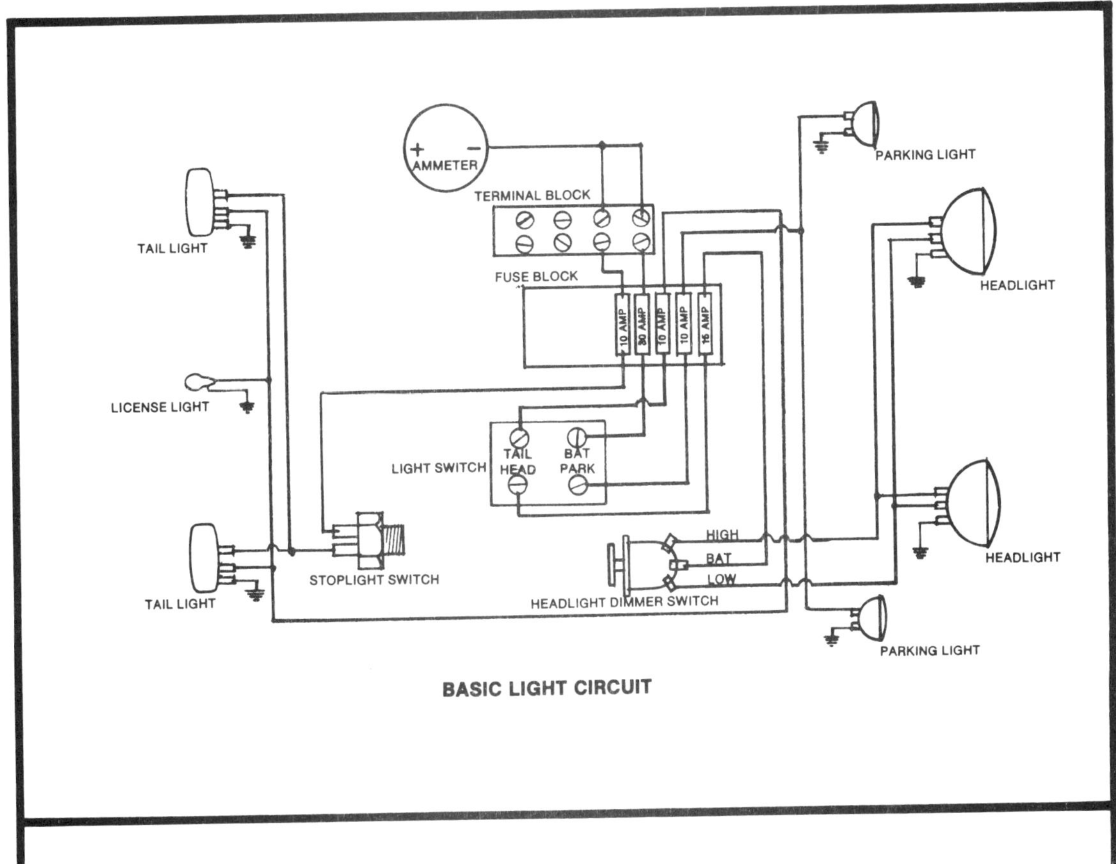

BASIC LIGHT CIRCUIT

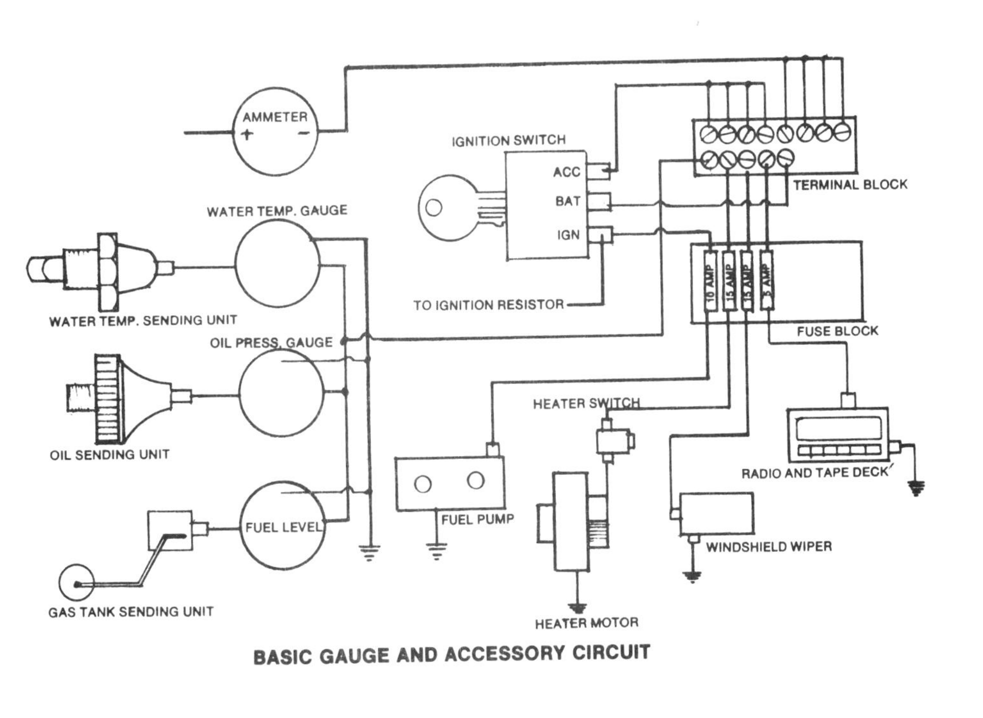

BASIC GAUGE AND ACCESSORY CIRCUIT

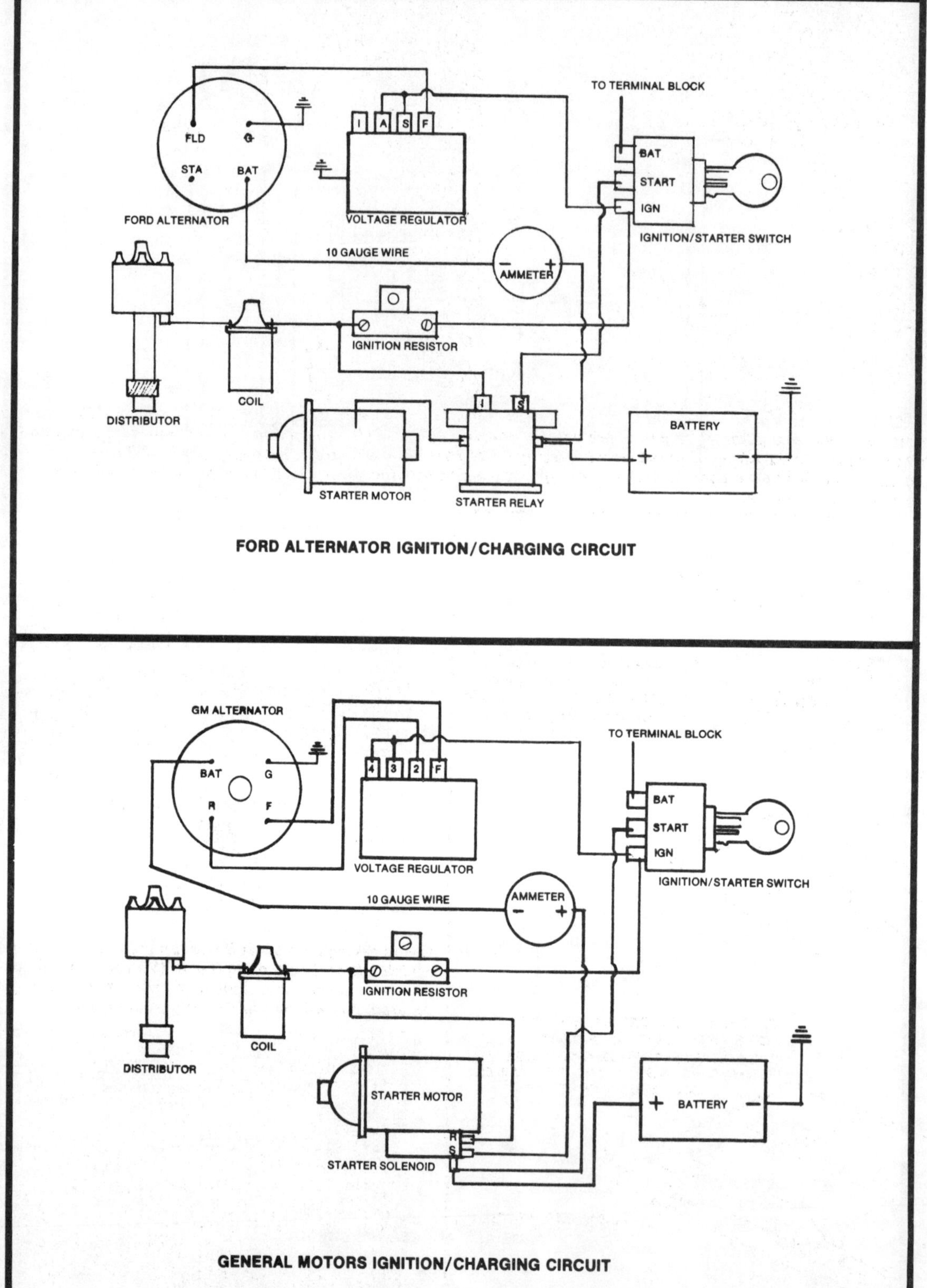
FLD
STA
BAT
FORD ALTERNATOR
TO TERMINAL BLOCK
I A S F
VOLTAGE REGULATOR
10 GAUGE WIRE
BAT
START
IGN
IGNITION/STARTER SWITCH
AMMETER
DISTRIBUTOR
COIL
IGNITION RESISTOR
STARTER MOTOR
I S
STARTER RELAY
BATTERY
FORD ALTERNATOR IGNITION/CHARGING CIRCUIT
GM ALTERNATOR
BAT
G
R
F
4 3 2 F
VOLTAGE REGULATOR
TO TERMINAL BLOCK
BAT
START
IGN
IGNITION/STARTER SWITCH
10 GAUGE WIRE
AMMETER
DISTRIBUTOR
COIL
IGNITION RESISTOR
STARTER MOTOR
R
S
STARTER SOLENOID
BATTERY
GENERAL MOTORS IGNITION/CHARGING CIRCUIT

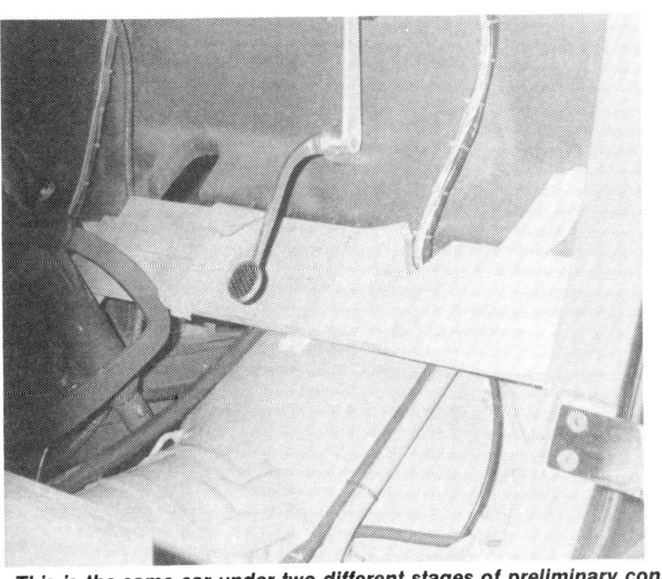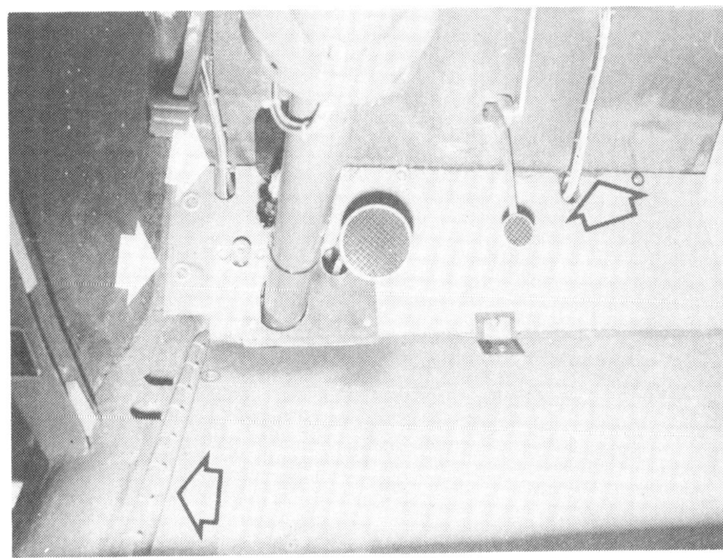

This is the same car under two different stages of preliminary construction involving the wiring. There's a lot to be learned here as changes evolved. Note how and where the wiring is routed, how the wiring related to the steering column is separated from the wiring related to the engine, how the toe board is attached to the body, how the mock-up of the toe board differs from the finished piece and finally how a main wire loom is routed from front to rear of the car.

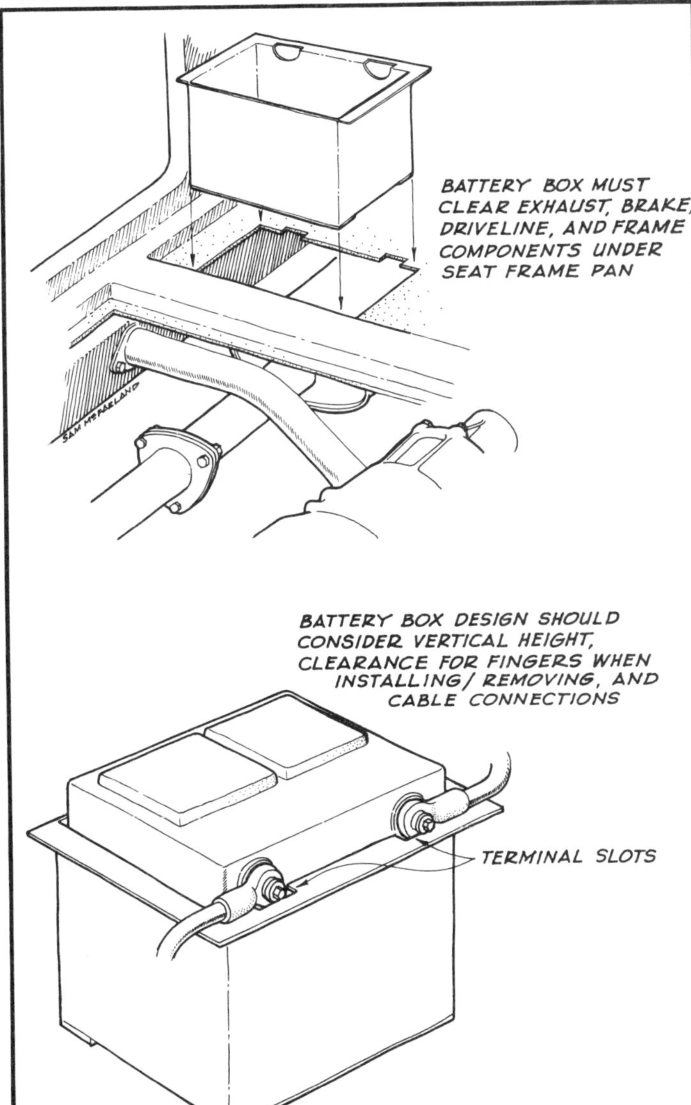

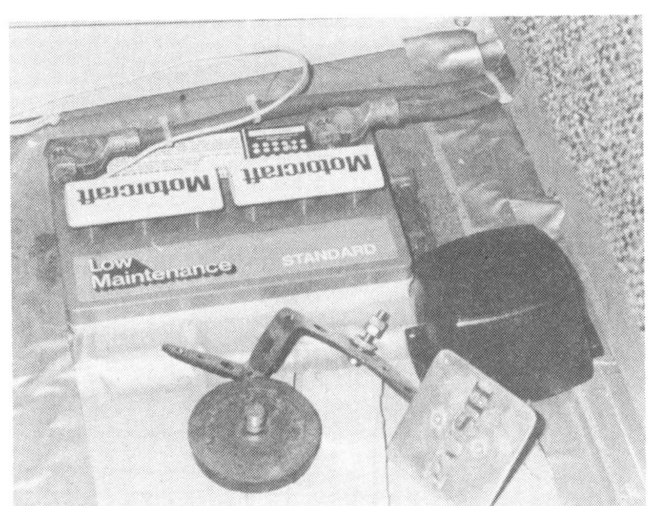

You're asking for trouble if you run the ground lead to any part of the vehicle other than the engine. In this case both leads go behind the seat back, under the floor and forward to the engine. If properly secured, this will be sanitary and trouble-free.

Here's a completed battery installation showing the master kill switch at the front of the seat riser. A hidden kill switch could prevent a fire or a theft. That's a pretty good investment.

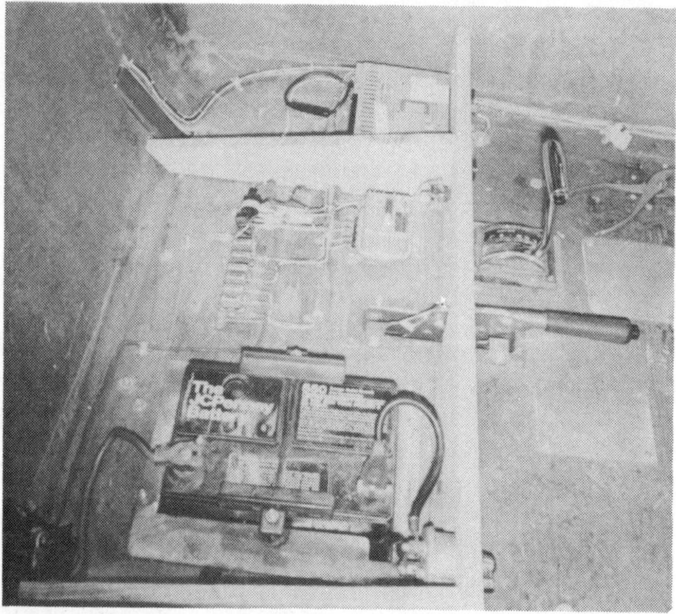

In this T roadster the major portion of the wiring is hidden beneath the seat. Replacing a fuse here is far easier than burying it far up in the cowl, but it also kills storage space under the seat.

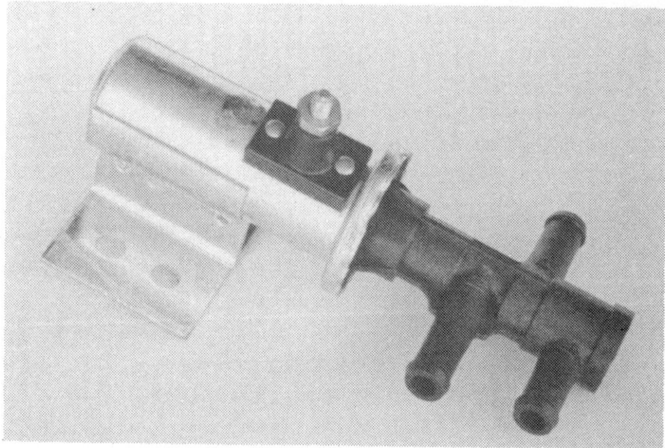

Ron Francis has this fuel selector valve for use with dual tanks. This is slick for A's with a fuel tank under each running board splash apron.

BATTERY MOUNT AND MASTER KILL SWITCH

Like a lot of other areas of construction in a street rod, battery placement is an area of compromise. Under-the-hood space is limited and a fiberglass firewall does not lend itself to the support of a heavy battery or other heavy pieces of hardware. Probably the most popular and practical placement of the battery is under the seat on the passenger side. The passenger side is chosen over the driver side due to the common placement of the master cylinder hardware under the driver. If this route is taken, keep in mind this has an effect on the construction of the seat bottom and how the exhaust will be routed around the battery. Side terminal batteries are neat for rods because the water level never has to be checked. Keep in mind the terminal location determines the construction of the battery box. Although this may seem as obvious as

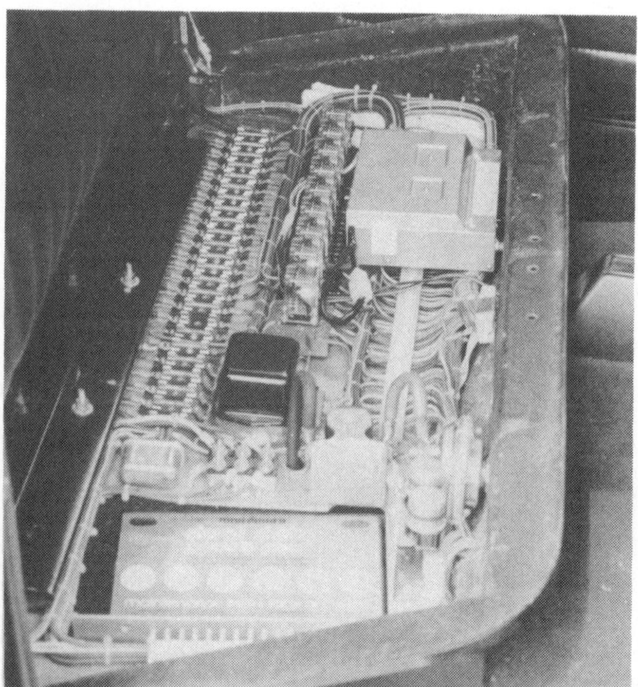

If you enjoy complexity for the sake of complexity you can have a ball in the wiring department.

Unless you enjoy a lot of headaches, keep the wiring simple and straightforward. Anchor it to keep it from moving around and protect it from water and road debris as best you can.

saying snow is cold, we know of at least two cases where a rod builder used an old junk, top-terminal battery to construct a battery box. When the boxes were finished, the side terminal battery would not fit.

Make your own battery cable. Use arc welding cable; most auto parts stores can supply both ends.

After letting the benefits sink in and living with the system in our own car, we highly recommend the use of a master kill switch installed in the positive side battery cable. In case of fire, this can save a car; and it is a considerable deterrant to theft. Ron Francis is one source for these switches. He has two types — one with a simple on/off handle, and the other with a removable key.

GROUND

We paint, polish, chrome and mount the engine and transmission in stock rubber encased mounting hardware, install all wiring, lights and instruments in a fiberglass body, spring for a high buck battery and then wonder why the engine won't turn over. Somewhere between the shuffle and jive and the turning of the key, we forgot that 12 volt electricity flows a lot better through a cast iron engine block than it does through a fiberglass body or rubber insulated engine mounts. When a fiberglass body is being used, special considerations must be given to a ground for all instruments, lights and accessories. Don't waste your time chasing electrical gremlins which seem so difficult, aloof and far away when a poor or non-existant ground is the culprit.

Route the negative side of the battery to an unpainted, easy-to-reach part of the engine. An under-the-hood inspection of a late model car will give you some ideas on the subject, and so will the instructions packed with a Ron Francis wiring harness.

INSTRUMENTS

Classic Instruments, AC, VDO and Stewart Warner are the most common names associated with street rod instrumentation. The selection, usage and placement of gauges in a street rod is primarily a matter of opinion. Cosmetic appeal is a very large part of the decision on what to use. We can tell you that electrically actuated gauges are more sanitary to install than mechanical gauges when it comes to water temperature and oil pressure.

Contemporary repro building leans toward one brand of instrument with all instruments installed in the dash. Classic Instruments makes a complete package designed for this application — including an electric speedometer. If you want to incorporate "nostalgia rod" thinking in your "repro rod," then you may be inclined to use a steering-column-mounted tach and under-the-dash-mounted mechanical oil pressure gauge.

Due to the immense popularity of the Classic Instruments, a few words are in order. Classic has two speedometers — one mechanical, one electrical. The electrical model is adjustable up to ten miles an hour either way from sixty miles per hour by simply turning an adjustment screw on the back of the speedometer head. If you are the type that continuously likes to tinker and change things — like rear axle ratio, and tires and wheels — this adjustment feature can save you world's of time. If the feature sounds neat, but an electrical speedometer does not, Classic can custom build a mechanical ratio adapter to correct their mechanical speedometer to read correctly with whatever tire/wheel and gear ratio combination you can come up with. Like other bits and pieces of paperwork associated with aftermarket equipment, the installation instructions of any instruments you buy should become part of a permanent notebook concerning the car.

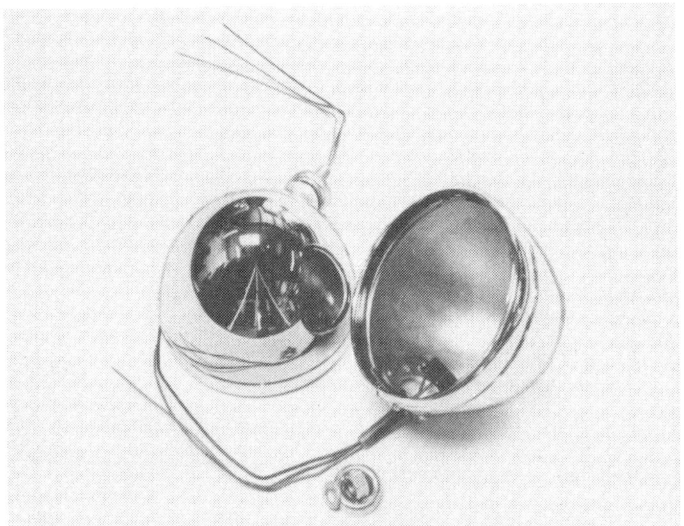

Depending on how far from stock appearance you want to go, there's a wide choice of headlight hardware including some of the very good quartz halogen lamps used in motorcycles.

If you have the talent to do this sort of machine work there is no limit to what can be created in the way of artistic and functional instrumentation packages.

(Left and Below Right) When it comes to instrumentation, you're on your own in a street rod. Here are two examples of starting with the same dash and different instruments and winding up with a lot of difference in appearance.

These are Classic Instruments installed in a fiberglass dash against a background of engine turned metal from Magoo. These instruments really are classic in appearance and are available in either chrome or brass. The accompanying wiring diagram from Classic shows just how simple these all-electric gauges (including the speedometer) are to hook up.

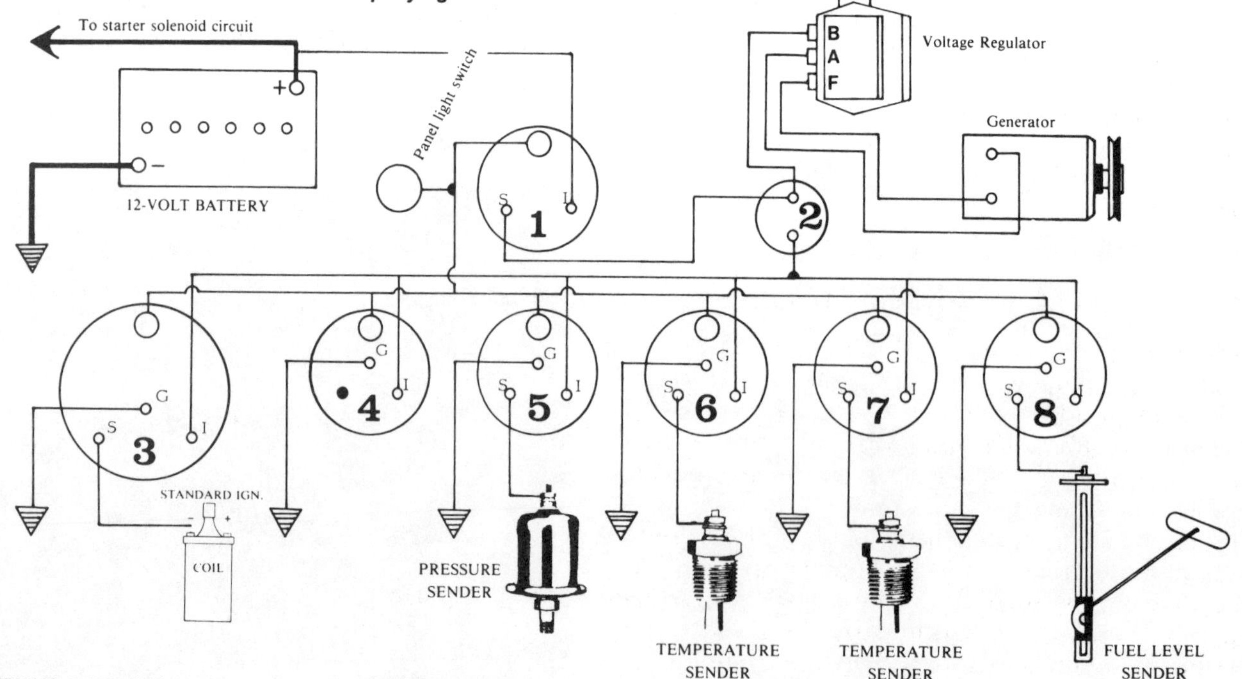

1 - DYNAMO GAUGE #866
2 - IGNITION SWITCH
3 - TACHOMETER #600 or 800
4 - BATTERY VOLTAGE GAUGE #830
5 - OIL PRESSURE GAUGE #881
6 - OIL TEMPERATURE GAUGE #828
7 - WATER TEMPERATURE GAUGE #826
8 - PETROL LEVEL GAUGE #810

IMPORTANT: This schematic shows gauges as they appear FROM THE BACK-SIDE and with the gauge light AT THE TOP!

Plumbing

The plumbing system includes all hoses, tanks and reservoirs as well as the small connecting hardware such as clamps, tubes and fittings. A great deal of the plumbing system deals with fluid storage and transportation in the cooling system, so a great deal of plumbing knowledge is contained in that chapter elsewhere in this book. This section will discuss the choice and fitting of plumbing lines.

STEEL BRAIDED HOSE

Stainless steel braided hose is both functional and practical for plumbing fuel, oil, braking and cooling systems while adding a touch of class to the race car. Besides looking nice, it is also nearly impossible to break, wear out, overheat, come apart or assemble wrong. It is also quite lightweight.

Inside the stainless steel fiber braiding, there are two types of inner liner material: Buna-N neoprene and teflon. For most all applications, the more flexible and less expensive neoprene liner is fine. However, for extreme pressure situations like the brake system, only teflon hose liner should be used because it resists expansion much better than neoprene.

When working with steel braided hose, it is important to understand the nomenclature of the size designations. The hose diameter is commonly measured in "dash numbers," for example, "dash two, dash five and dash

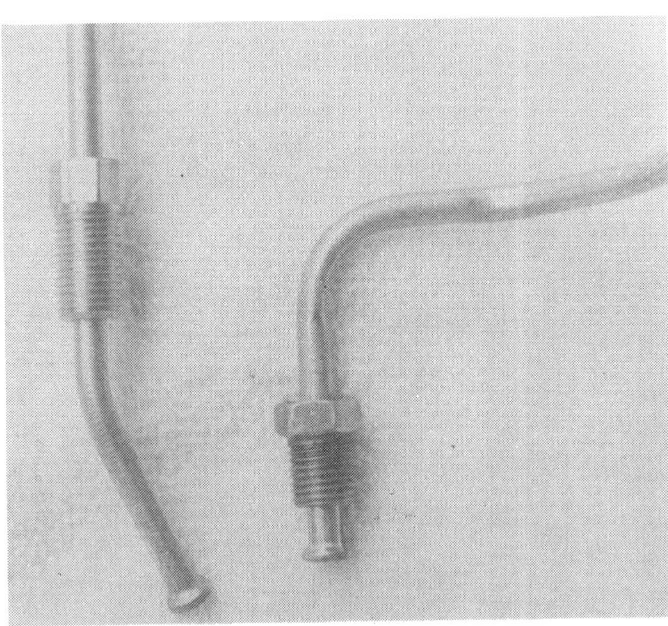

There is a wide variety of fittings available from numerous sources. Auto parts stores will have all you need to plumb a rod, but if you want the fancy stuff you'll have to turn to race car suppliers.

ten." The number following the dash is the number of sixteenths of an inch of hose diameter. For example, dash two (or -2) is 2/16 or 1/8-inch, dash five is 5/16-inch, and dash ten is 10/16 or 5/8-inch. The diameter this represents is the hose inside diameter equal to the inside diameter of a pipe with a similar outside diameter. For example, the dash six (3/8-inch) steel braided hose has the same inside diameter as a steel pipe designated 3/8-inch outside diameter. Consult the table for a complete listing of inside and outside diameters of all dash sizes.

HOSE ENDS

There are several brands and varieties of hose ends or threaded couplers for steel braided hose on the market. The four most popular types are distributed by Aeroquip, Edelbrock, Dave Russell's Race Car Parts and Earl's Supply. Get their catalogs to decide which fits your needs.

Threaded couplers with pipe thread should be coated with Permatex's Thread Sealant With Teflon (part #14) on the threads. It seals, lubricates, prevents corrosion and has an anti-sieze agent. If you use teflon tape, use care to prevent any tape ends or shreds from getting into the lines. It could pass on through the lines and particlly clog a part.

On AN threads, use oil, anti-seize or thread lube to prevent thread galling.

PLUMBING THE BRAKING SYSTEM

Plumbing the braking system involves the use of two kinds of lines, the solid lines and the flex lines at the wheels.

The solid brake tubing lines should be Bundyweld steel tubing in the 3/16-inch o.d. size, and not copper tubing. Copper tubing will not resist corrosion in use in the

STEEL BRAIDED HOSE SPECIFICATIONS

Dash Size	Liner	Hose i.d.	Hose o.d.	Maximum operating p.s.i.	Burst p.s.i.	Min. bend radius
— 3	neoprene	.156	.38	1,000	2,000	1.8"
— 3	teflon	.125	.25	1,500	12,000	1.5"
— 4	neoprene	.219	.44	1,000	2,000	2.0"
— 4	teflon	.190	.33	1,500	12,000	2.0"
— 5	neoprene	.281	.48	1,000	1,700	2.3"
— 6	neoprene	.344	.55	1,000	1,700	2.5"
— 8	neoprene	.438	.64	1,000	1,250	3.5"
—10	neoprene	.562	.80	1,000	1,250	4.0"
—12	neoprene	.688	.94	1,000	1,000	4.5"
—16	neoprene	.875	1.16	750	1,000	5.5"
—20	neoprene	1.14	1.44	500	750	8.0"
—24	neoprene	1.38	1.70	500	750	9.0"
—32	neoprene	1.78	2.02	200	700	12.5"

SUGGESTED USEAGE SIZE OF STEEL BRAIDED HOSE

Use	Liner	Size
Brake lines	teflon	— 3
Fuel lines	neoprene	— 8
Split log to carb fuel bowls	neoprene	— 6
Fuel pressure line	neoprene	— 3
Oil pressure gauge line	neoprene	— 3
Oil scavenge pump lines	neoprene	—10
Dry sump pressure line	neoprene	—12
Dry sump reservoir tank breather	neoprene	—12
Oil cooler lines	neoprene	— 8
Radiator hose	neoprene	—12

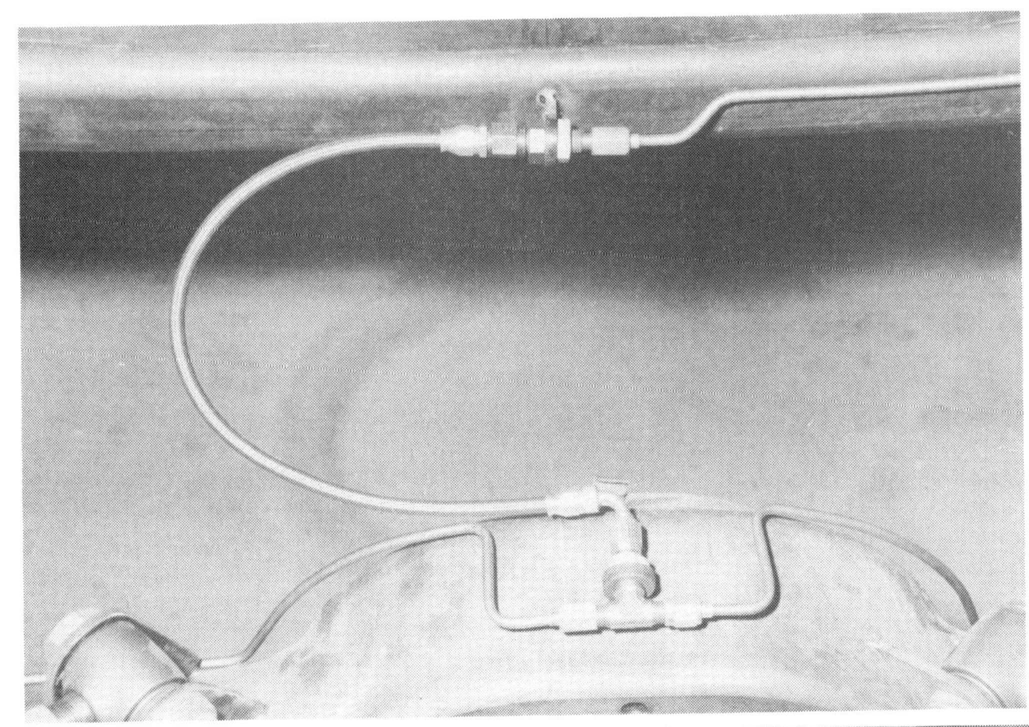

(Left and Below) Plumbing a street rod is one place you can be clearly creative. Note the very straightforward approach in both these examples and note the use of small tabs which accommodate bulkhead fittings which serve as anchoring points for the lines.

braking system, and will fatique and crack as well. The Bundyweld steel tubing is thick wall tubing which helps resist line expansion under pressure.

The layout of the solid tubing run from the master cylinder to each wheel should be carefully designed to prevent the presence of bends and loops which will trap air. In addition, to aid the bleeding procedure, the lines should run consistantly downhill from the master cylinder to the breaking junction at each wheel for the flex lines. This insures there is no place present in the lines to trap air.

At connecting points of the steel Bundyweld tubing, the fitting flares should be the standard automotive double flare. The double flare was designed to prevent tubing fractures at the flare. When cutting the steel tubing in preparation for the flare, use a tubing cutter, not a hack saw. The hack saw leaves rough edges and burrs on the tubing, and distorts the diameter and shape of the tubing end as well.

Never use a single flare on a steel brake line. Double flaring is a two-step procedure that takes a little getting used to, but forty-five minutes of practice takes care of that. If you encounter a run of tubing that wants to split during the flaring procedure, use a belt sander to remove the plating from the first inch of a length of tubing, then heat the end of the tubing to a cherry red; let it cool without quenching. This softens the tubing and makes it much more resistant to cracking.

Many rodders are now using stainless tubing brake lines. A parts house at the local airport can supply you with tubing, tools and fittings.

The steel brake lines should be secured to the chassis in several places along its run with rubber lined Adel clamps.

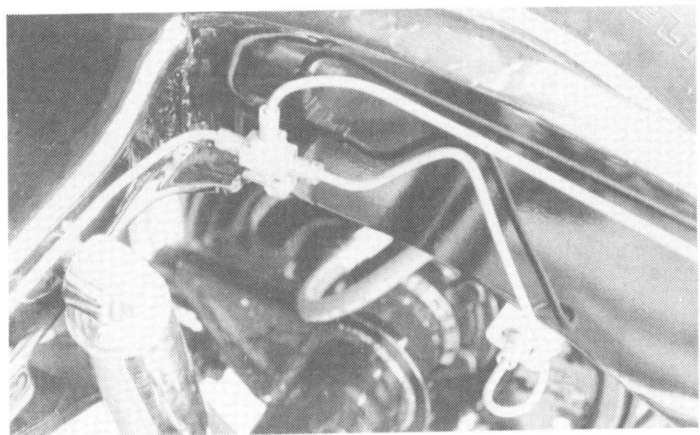

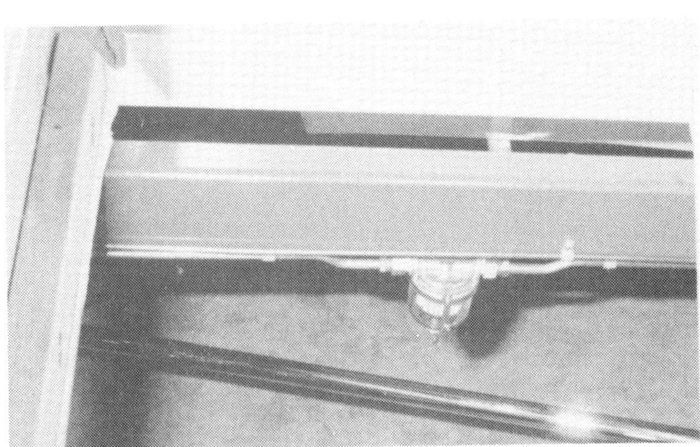

Here's still another example of cleanly routed plumbing. The brake line is routed at the very bottom edge of the inside of the frame rail while the fuel line is placed on the bottom of the frame near the inside edge. That fuel filter is bound to save someone some grief some day.

These aircraft-designed clamps prevent the transfer of vibration to the brake lines which could otherwise fracture the line material.

FLEX LINES

To prevent a loss of braking system rigidity and for extra safety, use steel braided teflon-lined hose for the flex lines. This is especially critical if disc brakes are being used on the car. If the steel braided hose is purchased as surplus material, be sure to determine if the hose liner material is teflon or neoprene. The neoprene material is not designed for use in high pressure applications like the braking system.

To determine the amount of flex line to use at each wheel, jack the car up and let the wheel rest a full droop. Then install a line of adequate length which is not tight. For front wheels, steer the wheel hard to one direction as well as setting at full droop to determine the maximum length of flex line required.

SILICONE BRAKE FLUID

This is the gee-whiz, racer brake fluid which is superior to ordinary brake fluid in every area you can name. We highly recommend the stuff for street rods for the simple reason it won't ruin the paint like ordinary brake fluid. That's reason enough.

BLEEDING THE BRAKING SYSTEM

To properly bleed any hydraulic braking system, the bleeder valves for the wheel cylinders *must* be at their highest point. On drum braking systems, simple system design and proper installation insures this will be done.

But on disc brake systems, it is possible to mount the calipers where the bleeder valves are not at their highest point. If this happens, air will collect in the fluid reservoirs above the level of the bleeder valve opening and remain trapped there during the bleeding process while pressurized fluid is passed through the bleeder valve.

If it should be necessary to mount a caliper in a position where the bleeder valve is not at the highest position, the caliper will have to be removed from its mounting bracket and held in the proper position while it is being bled. If this is done, be sure to place a wooden spacer between the brake pads equal to the approximate thickness of the rotor to prevent excessive piston travel during bleeding.

When first bleeding the braking system after the system's installation and assembly, open the connection between the solid steel lines and the flex lines to bleed the lines first. This prevents trapped air in the lines from being pushed into the wheel cylinders where it may be more difficult to remove. Then reassemble the flex line fittings and bleed the wheel cylinders. When bleeding the front wheels of the car, raise the rear end of the car. Conversely, raise the front of the car when bleeding the rear wheels.

GAS TANKS

There are several ways to solve "the gas tank problem" in a repro rod. The safest but most expensive way is to have a race car fuel cell built — a fuel cell assembly will consist of a steel container, a flexible puncture resistant bladder and safety foam which acts as a bladder. Firms like Aero Tec Laboratories have standard cells and can custom build most any shape you can dream up. There are endless combinations of check valves, filler openings and associated plumbing which can be used to solve all manner of fitment problems and general plumbing. You can even

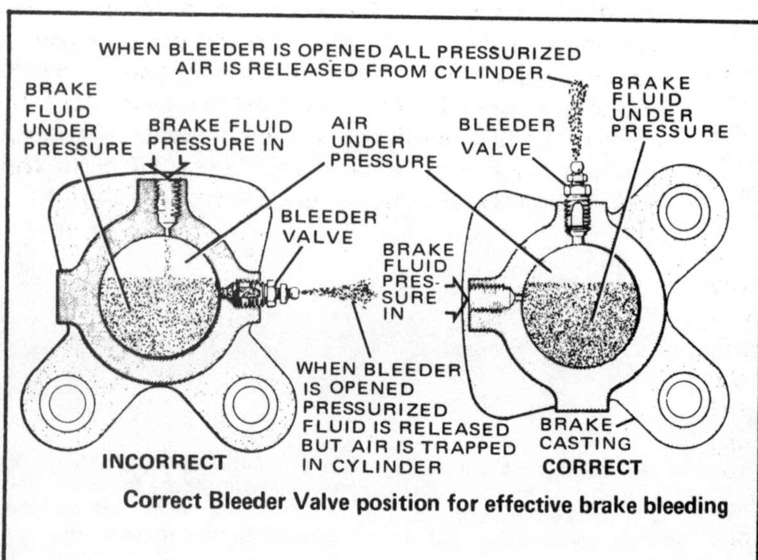

Correct Bleeder Valve position for effective brake bleeding

GM dealership parts counters sell this replacement element fuel filter which is easy to mount to a frame rail.

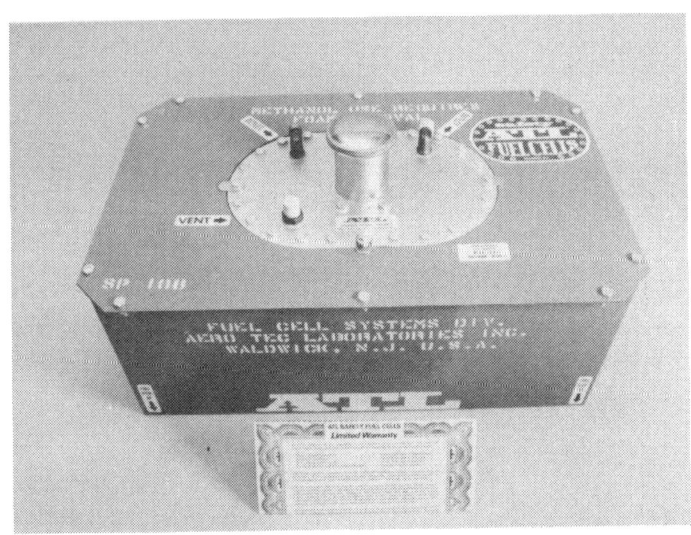

The top of the line device for holding fuel in any kind of vehicle is a foam-filled bladder in a fuel cell container such as this one from ATL.

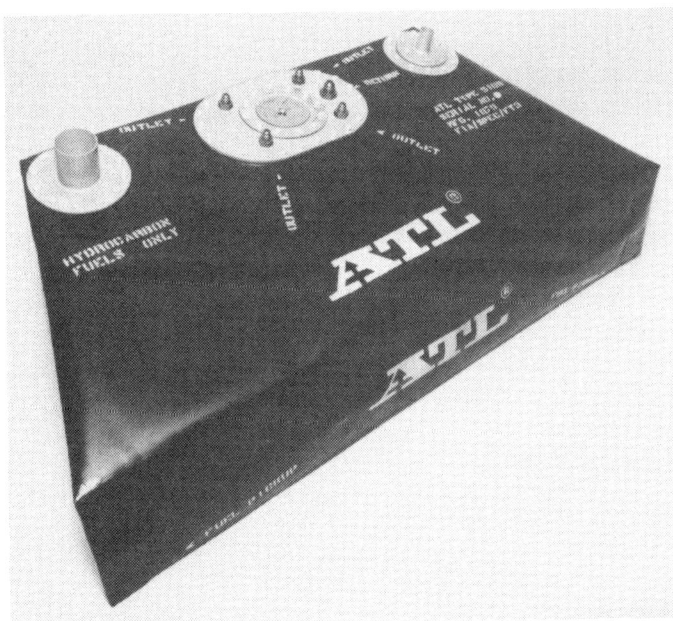

The bladder itself is made of a rubberized fabric of the same type used to make bullet-proof vests. A fuel cell such as this one should always be placed in a metal fuel cell container.

get a fuel level gauge designed especially for use with a fuel cell. Count on a 15 gallon fuel cell costing you a minimum of five hundred dollars.

A fuel tank can be custom built by a sheet metal shop in most any small town if you will supply them with dimensions and the placement of fittings.

Magoo has a "standard" tank he fabricates for use in Model A repro rods and is designed for maximum leg room in a car with a rumble seat and maximum storage in a car with a trunk.

The fourth alternative for a gas tank is to spend a lot of time in a wrecking yard looking for what you want. Rectangular, 12 to 15 gallon tanks are available that are suitable for street rod use. Any used tank should be throughly flushed with water and then taken to a local radiator shop to have a fuel level sender unit installed which is compatible with the fuel gauge you will be using. (If you use a Classic gas gauge, then use a Classic sender assembly.) After the correct collar for the sender unit has been soldered in the tank, remove all of the sender assembly, plug the fuel line fitting and sender unit hole and then pour about a quart of muratic acid into the tank and slosh it around thoroughly to remove rust and assorted corrosion. Drain the acid out, let the tank dry thoroughly on the inside and then pour in a quart of industrial tank slushing compound and again seal the tank and rotate it by hand until all surfaces of the inside of the tank are coated. The slushing compound prevents further rusting and will seal pin hole leaks. One such slushing compound is marketed through Fuller-O'Brien paint stores. There are others. (Some radiator shops can boil out a gas tank and pressure check it for leaks. This is a solid plan.)

When installing any gas gauge sending unit in any gas tank, DO NOT USE ANY KIND OF SILICONE SEALANT AT ANY POINT IN THE INSTALLATION! From personal experience we can tell you that silicone sealant and gas do not mix. The sealant balls up and plugs the fuel line and just when you think you have seen the last of it — it happens again.

FUEL SHUT-OFF VALVE

Like the electrical master kill switch between the battery and the rest of the electrical system, a manual fuel shut-off valve between fuel tank and fuel pump is a good deal. It can save a lot of grief and it can also serve as an anti-theft deterrent. Mount the valve so you have reasonable access to it. Turn it on and off after you've finished the car to be very comfortable about how to locate it in the dark. Just for your own education turn the valve to the OFF position, start the car and drive it until it runs out of gas. How far did you get? Keep it in mind. This information could come in very handy some day!

Paint

PAINTING THE FRAME

There are a couple of ways to approach painting the frame — do you want it to be perfect, or do you want it to be nice, and have plenty of paint on it so it won't rust? Let's take the "perfect" approach first.

After all welding, drilling and other work has been completed, take the bare frame outside, turn the hose on it and leave it outside long enough to get a light dusting of rust on it. Now haul the frame to whoever does sand blasting in your area and tell them to remove all of the rust. Human nature being what it is, the sand blaster might just do a half-way job of sand blasting if you take him a rust-free frame. If you take him a rusty frame and tell him you have to have all of the rust removed because you are building the wildest rod in the world, then he'll have to remove all the rust 'cause he knows it will be very easy for you to check it out when you come to pick up the frame.

Be ready to start spraying primer as soon as you get the frame back. Lay on several light to medium coats of a good red oxide primer — following the directions to the letter. If you are going after the "perfect" paint job (show quality) on a frame, spray on Featherfill which is essentially a catalyst type "spray-on bondo." Follow the directions and spend hours with a rubber block and #360 grit sand paper (dry only) and work with the frame until it

is as "perfect" as you want it. Then seal with something like DuPont #2129 sealer; go over it with a tack rag and paint with DuPont Centari or the equivalent.

If you are not after "perfect," leave out the Featherfill and all the hours of sanding, but do everything else.

Most top-ranked rod builders feel that powder paint is the ultimate finish for a street rod frame and for much of the running gear. Powder painting must be done in a commercial establishment specifically set up to do it. The paint is dry — almost like talcum powder. After the component is painted it is then hung in an oven and baked. The result is a very tough, durable glossy finish that is imbedded in the metal. No primer is involved. The finish is not affected by gasoline, lacquer thinner or brake fluid. Ask other street rodders in your area where powder painting is done near you or search the yellow pages of a large metropolitan area.

PAINTING THE BODY

As car enthusiasts know, there are paint jobs and there are paint jobs. Secondly, if you talk to five painters about putting a quality paint job on a repro rod, you'll get five opinions on how it should be done. Don't consider that bad news because all five might be right. The procedure

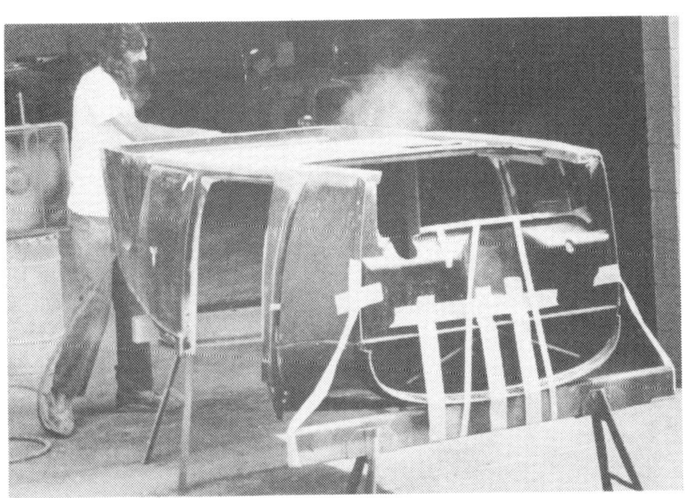

Paul Bonant at Magoo's figures the underside of the body should look as good as the top — and when he finishes painting a rod, it does.

Morton Eliminator is an excellent primer, surfacer, sealer for any fiberglass part. This stuff can be difficult to find in some areas of the country, but it's worth the search.

outlined here is the one currently used by Paul Bonant who paints the cars built by Magoo's Auto. The quality is simply superb.

The body prepping begins with sanding all individual body panels with 80 grit sand paper backed by a long sanding block. The purpose of this is to make all surfaces true. The combination of the long block and the heavy grit knock down the "mountains" on the surface in a hurry. Once the high spots have been taken down, the low spots can be filled in with ordinary plastic body filler. After this dries the surface can be finally trued with the 80 grit and long sanding block. For a full fendered Model A roadster with a new fiberglass body in very good condition, count on this first step to take a minimum of 10 hours. If there is a problem door and deck lid and one fender with surface irregularities, the job can easily stretch to three eight-hour days for a professional and a week for the amateur who achieves the same results. When all surfaces are true, all parts are then washed with Prep Sol.

The second major step in preparing a fiberglass body for painting is to spray all individual parts with a quality primer/surfacer/sealer which is mixed with a catalyst. Use Morton Paints Eliminator or equivalent. This is a critical step. Follow the directions on the can to the letter. After drying, the primer/surfacer/sealer should be lightly fogged with a guide coat of black lacquer. You want just enough black lacquer on the surface to serve as a reference guide when you begin sanding it off.

All parts can now be sanded with 320 grit wet or dry paper using a solution of water and mild detergent. Still using the previous example of a full fendered Model A roadster, count on this wet sanding procedure taking two full days. You should also know that because the primer/surfacer/sealant is very hard, you'll use up a lot of sandpaper. As each body panel is finished, it should be washed thoroughly with clean water and left to air dry.

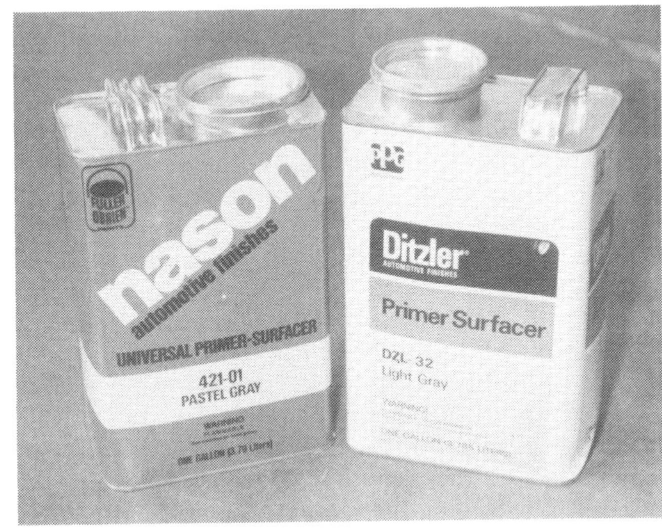

There are all sorts of primer surfacer. The price is normally a good guide to the quality — buy the best you can find.

The panels can now be shot with a good quality acrylic primer surfacer. After this is dry, another guide coat of black lacquer can be applied. Each panel should now be sanded with wet or dry 400 grit paper or dry 320 grit pre-cut paper. All panels must now be washed with Prep Sol.

Each panel may now be painted with nitro cellulose or acrylic lacquer of your choice. Follow instructions on the can to the letter. You should know that for the amateur painter shooting in a garage or driveway, the nitro

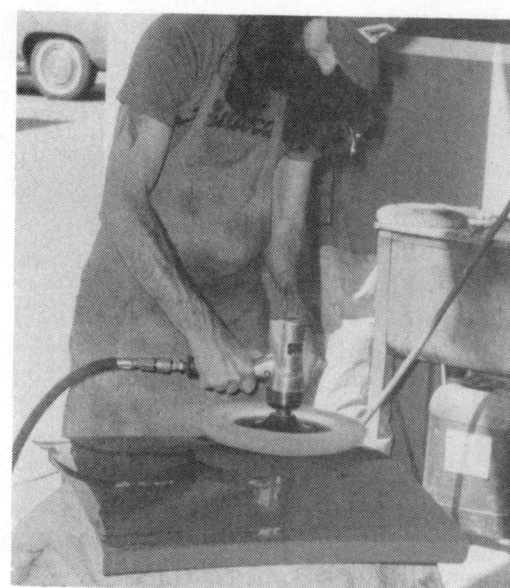

(Left and Above) The final step before assembling the panels into a complete car body is apply rubbing compound either by hand or machine and follow that with polishing compound.

cellulose is somewhat more forgiving. Make sure the floor of the surrounding area is wetted down before painting, that the air is still and that the compressor has the pressure capacity required (this will be stated on the paint can). For this first application of color, spray on at least four heavy coats and let them dry for four to six days. At the end of this period of time the panels should be color sanded with dry 320 grit or wet or dry 400 grit. Wash all panels with Prep Sol and reshoot another four or five coats of color. Let this second shooting of color dry for four to six days and final color sand with ultra fine (1000-1200 grit) paper and soapy water.

The panels are now ready to be machine or hand rubbed with rubbing compound followed with polishing compound. The final step at Magoo's is to go over each panel with Mequiar's Sealer and Reseal Glaze (MGH-7) which keeps the paint looking wet. The body panels may now be assembled.

If the underside of the body is to be shot in color, then this should be done before shooting the top side — and the bottom side should be painted with an acrylic enamel (such as DuPont Centari) and allowed to set for a week before disturbing.

Any painter worthy of the title will tell you that a good paint job is 95 percent dependant on all preparation before the paint is applied. Without exaggeration, several thousand dollars can be saved on a repro rod project by prepping the body and painting it yourself. Several hundred hours of labor are involved, but the fact remains that an amateur can compete directly with a professional in terms of a quality paint job if no short cuts are taken.

An optional move to consider is to talk the job over with a car painter with a good reputation. It may be possible to do all preparation and even color sanding, and let them apply the color if you feel that particular aspect of the job is beyond your capabilities.

This product from Meguiar's keeps the paint looking wet. This is a lot more fun than driving around in the rain all of the time to achieve the same effect.

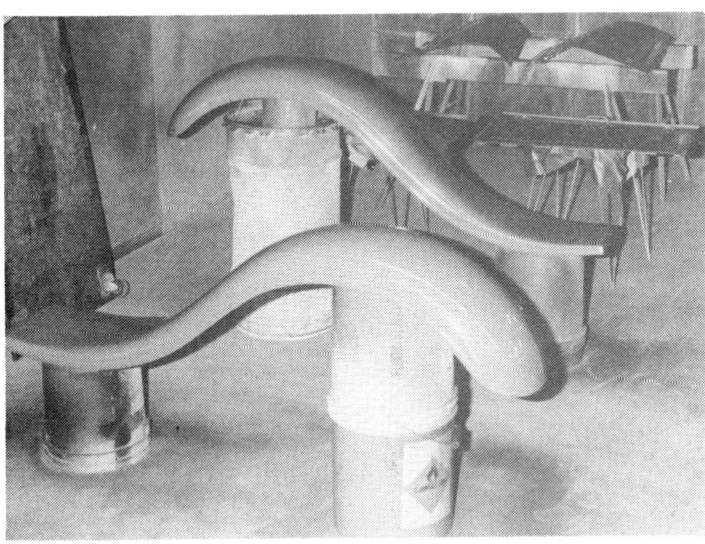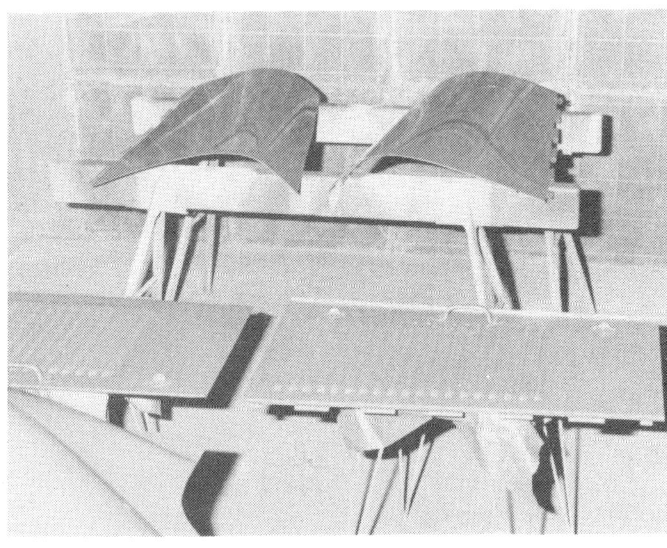

The only right way to paint a rod is piece by piece. If you're stumped as to how to prop up the fenders, take a hint from Paul and start collecting cans and saw horses.

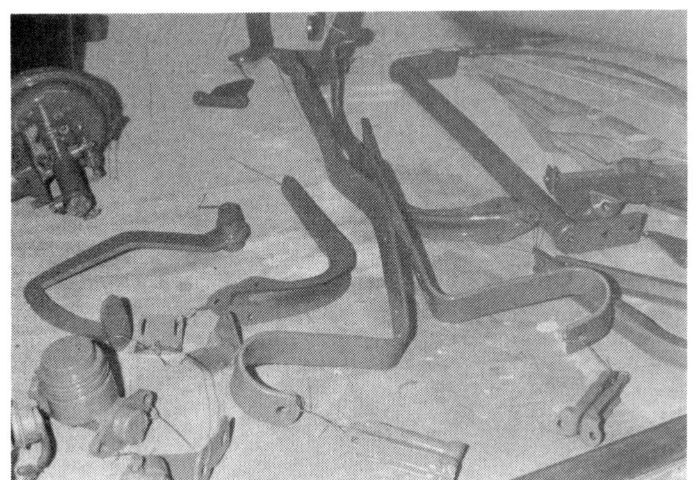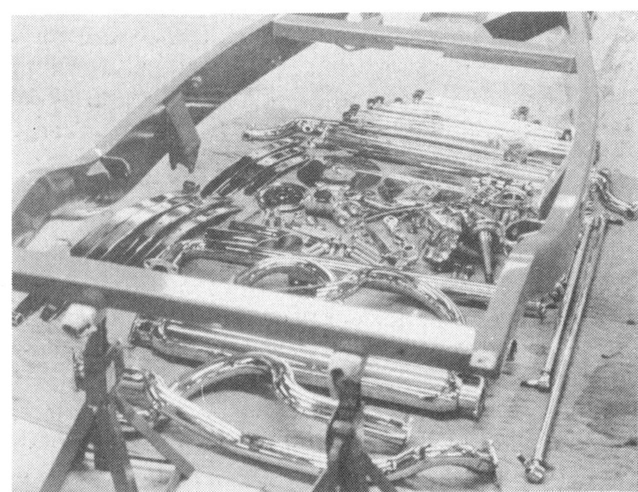

All of the scores of pieces of hardware which are essential to a rod should be individually painted or plated prior to final assembly. This is an excellent test of one's organizational abilities.

Upholstery and Seat Construction

Quality upholstery is an acquired skill demanding of time and patience. Obviously it is a skill which can be taught — but not in one chapter of a book. If you do not upholster, it is important to be able to communicate with someone that will be doing it. To aid the person doing the work, you get very specific about what you do and do not want.

The foundation for upholstery in a repro rod is the wood work. Correctly done, this step is exacting and tedious. If someone other than the upholsterer does the woodwork, much of the effort can be wasted if the woodworker does one thing and the upholsterer wants something else. If the upholsterer has to rework the wood, then you will be paying for something twice. If the upholsterer accepts the woodwork but doesn't like it, you most likely will get an upholstery job he doesn't particularly care for — and you won't either.

Before any wood is cut or any stitch is taken, you must first decide what you want — pleats, button tufting, plain, cloth, leather, vinyl, or a combination, and what color(s). If you are building the car primarily for the show circuit, you most likely will want something other than what you would prefer for every day street use. Look at a lot of cars and when you see something you like, ask a lot of questions.

A separate pattern should be made for every panel and then transferred to door skin or panel board. If the upholsterer will work with you on this, several hundred dollars can be saved just by making the patterns yourself.

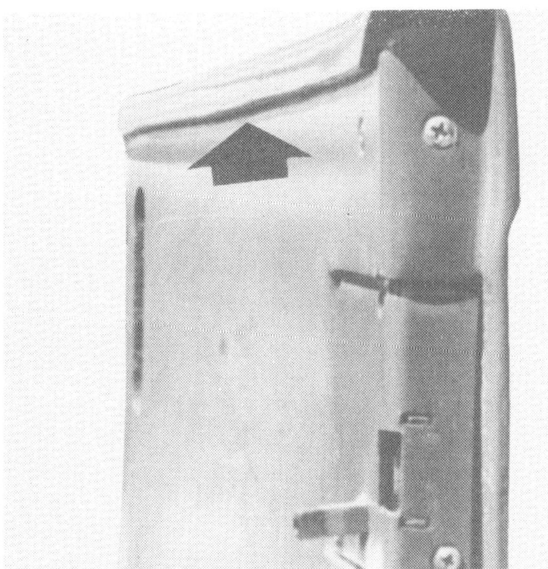

On a stock Model A, the door panel was tucked under this lip. On most reproduction bodies this can't be done. Does the upholsterer want to butt his panel against the lip or go over it? This is an important consideration when making the pattern.

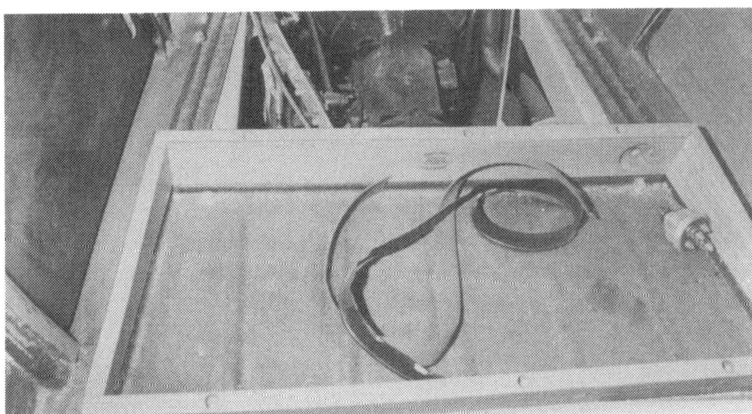

Before plunging into the upholstery project take a look at the seat risers of a number of cars the same make the vintage of yours. In the first place you'll want to be working with a wood seat riser — not a fiberglass unit. Secondly you may want it shorter than stock if you are taller than a telephone pole.

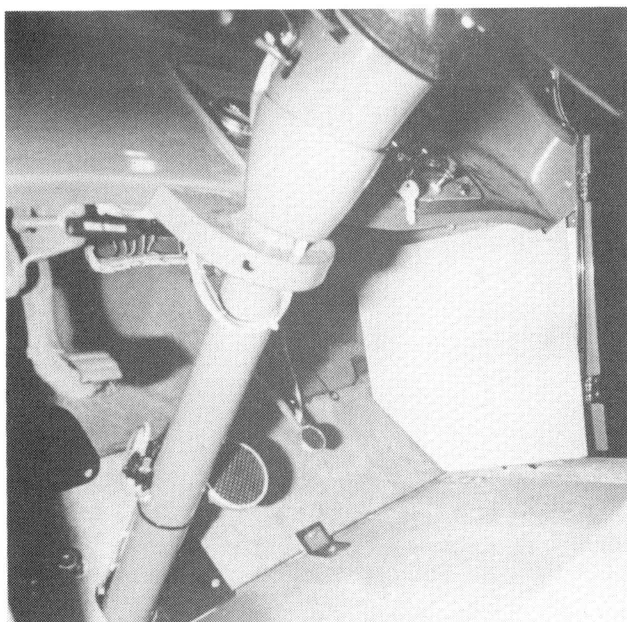

Hand fitting each panel in place is time consuming and often frustrating, but keep in mind that if it doesn't fit at this stage the problems will just get progressively worse. This is also the time to start thinking about how the panels will be attached and subsequently removed. You can't really appreciate the latter feature until you have to work on a rod having all upholstery glued in place, leaving no method of access behind the panel.

THE WOOD FOUNDATION

Now to the wood. The first step is to make cardboard patterns for everything — with a full understanding of what the upholsterer wants in terms of dimensions and material. For instance, one upholsterer may want the door panel wood to extend to the top of a roadster door while another may want it go only to a certain lip. One may want to upholster over waterproof panel board and someone else may prefer 1/16-inch mahogany door skin. Not so incidentally, door skin is far more durable and less resistant to warping than panel board. For the ultimate in resistance to warping and panel deterioration, coat the door skin with resin after it is cut to size.

For floor boards, seat bottoms and backs, and any other area where strength is important, one material to consider is ½-inch plywood covered on one side with a medium density overlay. This is cabinet builders material — and is not readily available in every lumber yard. The overlay gives a much smoother and attractive finish than normal plywood. In the case of the floorboards or other panels exposed to the elements, they should be coated with Eliminator or resin to completely seal the pores of the wood from moisture.

Three-inch polyfoam can be used as cushion material for the seat bottom and back. It is comfortable and far more space efficient and workable than coil springs and padding.

CHOOSING MATERIALS

Spend time with the upholsterer talking about material. Don't be afraid to ask about differences in cost, durability and resistance to fading. For instance, as this is written, leather costs about forty bucks a yard before it's installed. There may be some descrepancy between what you want and what you can afford. Mixing a cloth with vinyl may give you what you want in terms of feel and looks, but you should know that different materials fade at different rates. That could be an important consideration if you paln to keep the car for a long time — and the car is a roadster with no top.

You should know that Naugahyde is a brand name of

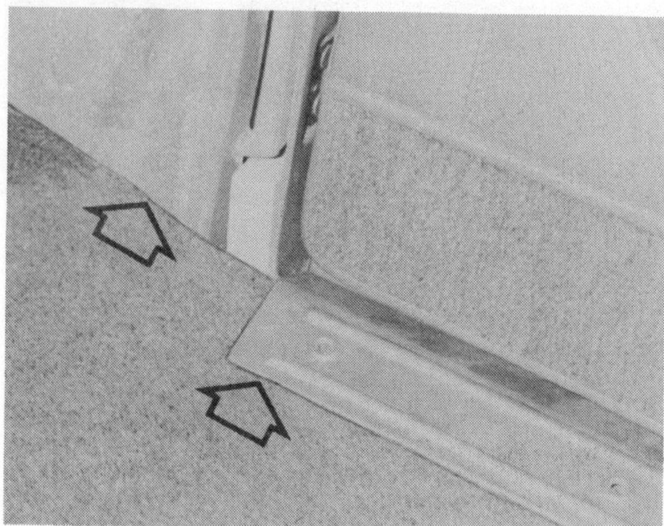

For best appearance and practicality, carpeting should be run under kick panels. Use sill plates and glue to keep the carpet from shifting. Most of the wear and tear to street rod carpeting comes from getting in and out of the car. From the very first day you get in the car you should make a practice of making contact with as little of the carpet and upholstery as possible when entering.

Spend all the time you need in constructing, fitting and sealing the floorboards around wiring, pedals and steering column. When constructing any major portion of a street rod — such as the floorboards — keep asking yourself all sorts of dumb questions ... like, can I remove and replace the master cylinder, steering column and battery without removing the floorboard? What about transmission and/or shift linkage? What about adding brake fluid to the master cylinder? A little extra time and effort when the car is going together can save a lot of time later on.

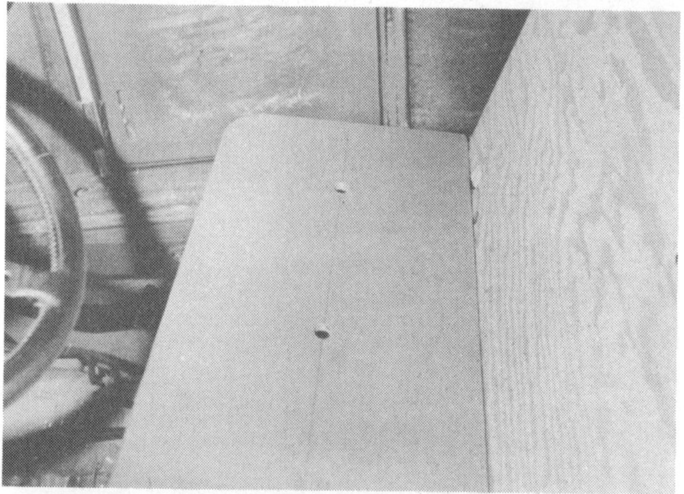

The base of this seat cushion shows the finish of the medium density overlay plywood mentioned in the text. Seat cushion in a street rod should always be easily removable. Plan on it from the beginning. There's lot of space for storage under that cushion.

vinyl upholstery material and is among the best there is. There are many qualities of vinyl and much of the quality is quite a bit less than the best.

A good upholsterer will take care of most (if not all) of the carpeting before starting on kickpanels and seats so the carpeting will be run under kickpanels. By doing it this way any shrinkage will still be hidden if it occurs. You should be prepared to talk about carpet as well as seat and door panel material. Cut pile carpet such as the kind used on more expensive European and American cars is more durable than loop. If a loop is snagged, an entire line of loops can be pulled from the carpet because they are all connected. Such is not the case with cut pile.

Be wary of an upholsterer who has yards and yards of material on hand. Is he trying to unload some inventory on you and make more money on the service he performs also? Long before your car is ready to be upholstered, you should be attending rod runs in your area and asking questions about talent in your area. If you see upholstery you like, ask who did it; the same goes for upholstery you don't like. A quality upholstery job in a street rod can amount to a sizeable investment. Do a lot of looking and ask a lot of questions before you boil the entire upholstery situation down to what you finally wind up with.

Finally, if you are paying a lot of money for a top-of-the-line upholstery and carpeting job, ask if you will be getting handroll binding. You should. Handroll binding is a two-step process where the binding material is bound upside down on the upside of the carpet, rolled over the edge and stitched again from the backside. To the keen eye, there's a world of difference between top-of-the-line upholstery job and a run-of-the-mill job. Good luck.

ROADSTER TOPS

Like all the rest of reproduction parts you'll encounter when building a repro rod, you'll discover there are great variances in the quality of top bows and top construction. Probably the best quality top bows come from Gates Products in Upland, California (see Suppliers List). Gates offers the top irons in painted and chromed versions and with or without the steam-bent oak attached. The chromed version of the irons for rods feature stainless

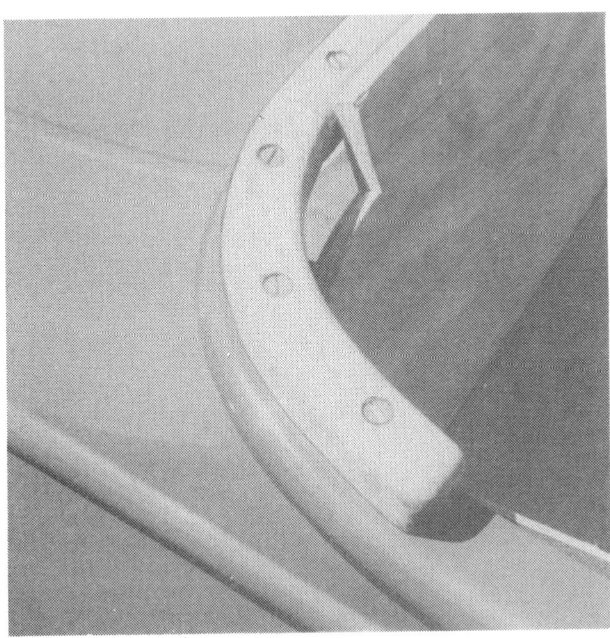 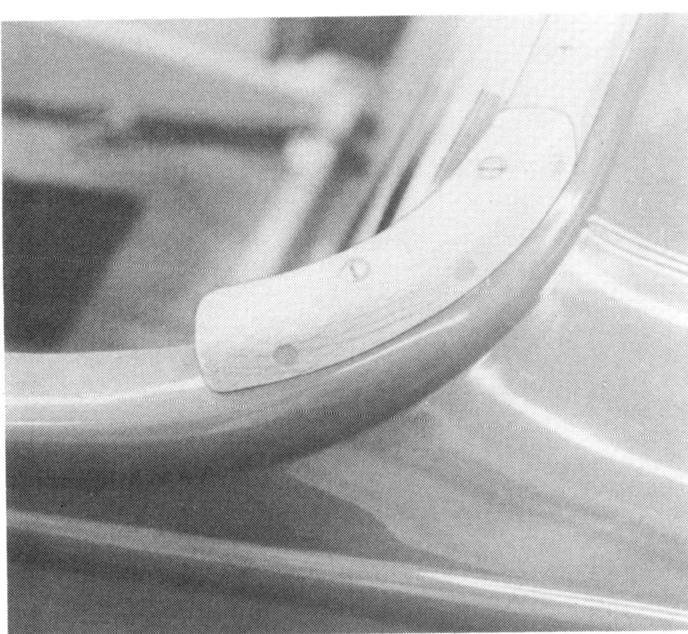

This is the kind of problem encountered when more than one cook is stirring the stew on a street rod. In this case the guy doing the woodwork wrapped the wood at the back of the seat further around the quarter panel (three screws) than the upholsterer wanted it. The result was a rework (two screws).

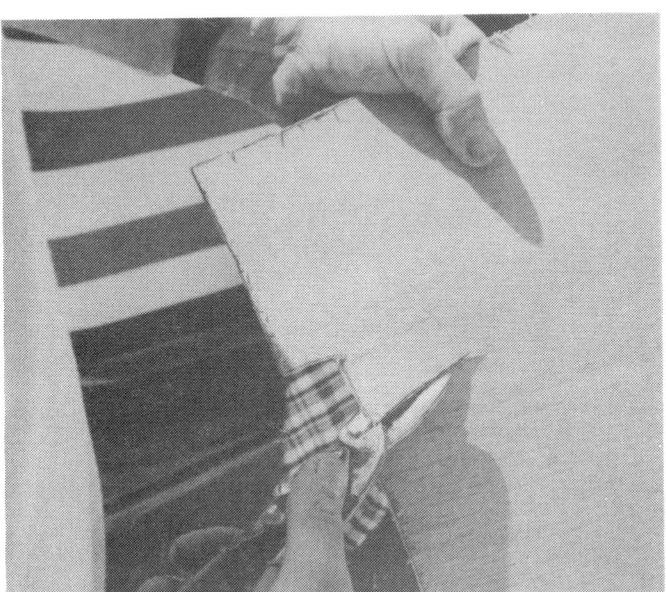

Door skin is readily available in large lumber yards and is far more durable than the panel board found in upholstery shops. It is far more difficult to stitch through and for that reason many upholsterers avoid the material which is easily cut with shears, band saw, saber saw or sharp knife.

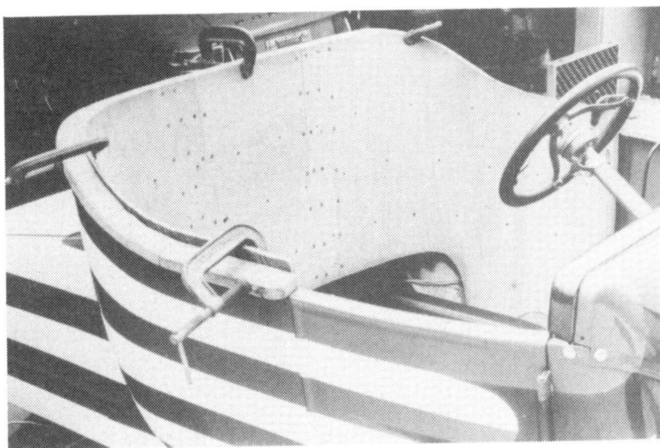

If a seat former such as this is built for a basis for upholstery, construction is best accomplished by soaking door skin in water, then clamping it in place. Build up three layers of wood with a water-soluble glue between the layers. Leave final trimming until all layers are together.

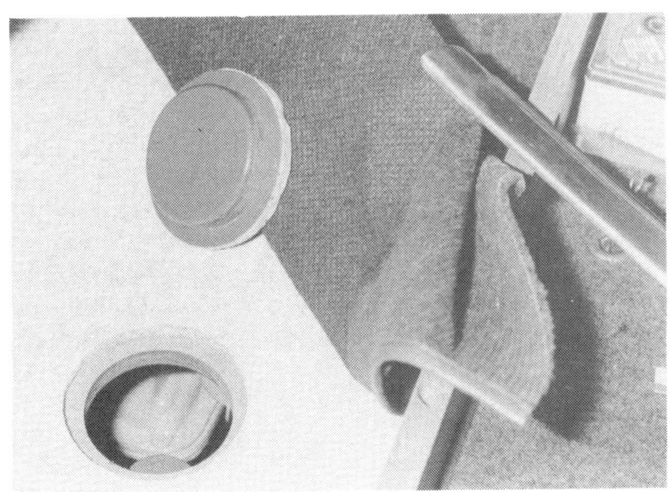

(Right) This is the sort of planning that makes a top-notch street rod. In this case an access plug has been built into the floorboard for filling the master cylinder. In this area the carpet has not been glued to the rebond; clamping is accomplished via the sill plate.

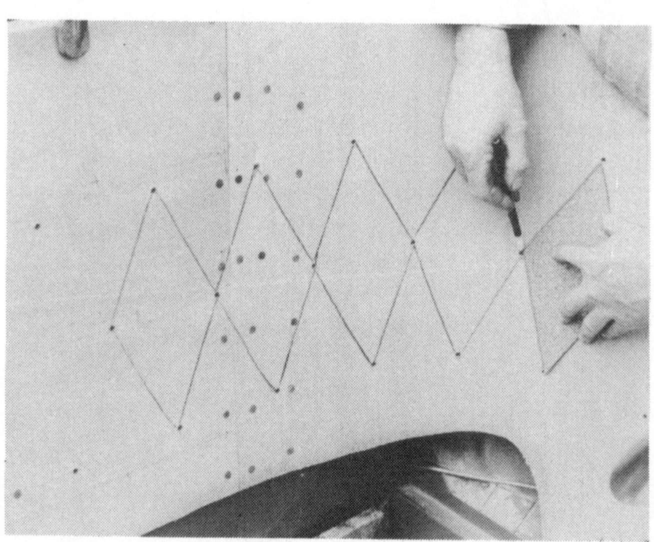

(Above and Right) Regardless of the car or how it is being upholstered, layout is of primary importance. If you cannot communicate with the upholsterer in this regard you will be an instant back seat passenger in the entire upholstery project.

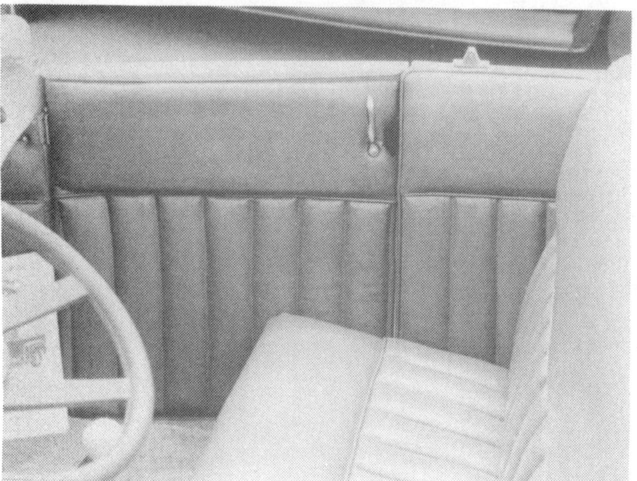

In these four examples of door panel upholstery there is a considerable amount of variance. Overlooking the obvious design differences, note the wrinkles or lack thereof; note the door jamb showing on some cars and not on others. Note the vast differences of design available on panels almost identical in size.

rivets and washers at the pivot points for a lifetime of rust-free operation. Neil Gates has top assemblies for '28 through '34 Ford roadsters and phaetons.

Even after you have a set of Gates top irons and bows, an upholsterer will be needed to complete the top — and after you start looking at cars and photos, you'll find there is a considerable variation as to what can be achieved. Regardless of who stitches your top, you'll probably be more pleased with the end result if you communicate with the craftsman in some detail as to exactly what you want.

Here's a very clean top installation on a '34 Model 40 roadster.

(Left) The Gates top bow assemblies for rods are works of art. All of the iron pieces are polished and then highly chromed before being assembled with stainless steel rivets. The steam bent oak bows are finished like fine furniture with three coats of polyurethane finish.

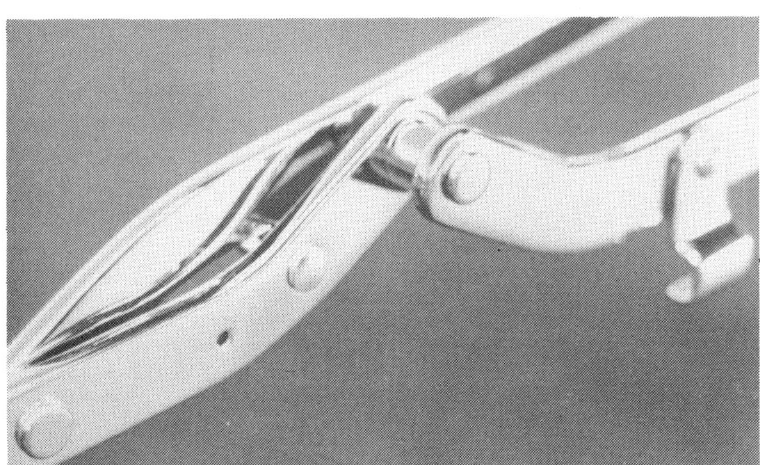

Some of the best chrome work we've ever seen is used on the Gates iron assemblies. Neil also does custom top iron assemblies when time permits.

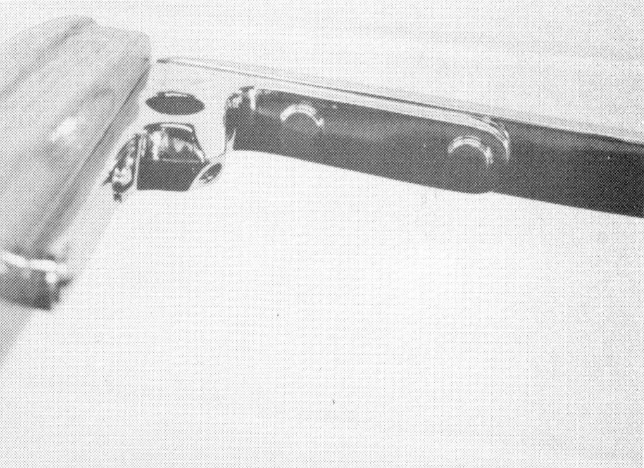

The corner bracket that drops over the windshield post is a polished brass casting that is chromed before being attached to the rest of the top iron assembly.

Note the difference in these '32 roadster tops. If you like one better than the other then you should make your ideas known to the upholsterer before he goes to work on your top.

Here's another '32 roadster with a chopped top. Depending on how large you are, this may not be such a good idea. If you do want a chopped top, ask if you can sit in several cars that have them before making the decision.

Here are two examples of the rear window panel folding up. This makes for very good visibility to the rear when the panel is in the up position.

Here's a very nice custom-built roadster top that will not fold, but can be removed from the car. On a rod run, the top provides protection while enroute to the meet and can then be taken off and stored in a motel room. The center top bow is relatively high which provides a maximum amount of head room.

Two different approaches to the rumble seat back show variances in material, design and comfort. You should understand before buying your body that there is precious little room for storage in a rumble seat roadster.

The Finishing Touches

A rod, a street rod, a repro rod — no matter how many hands had a part in building it — is, in the end, an individual expression in a relatively narrow band of activity. Some expressions come off so well they are regarded by the beholder as an art form. Although much of the success of such a car can be traced back to the very initial planning stages, the final "finishing touches" are very important. This chapter could have been ten times as long as it is — a nit-picking look at all details of rod building. Much of this has already been thoroughly covered in a companion book from Steve Smith Autosports entitled "Street Rod Building Skills." That would be overkill. What we really want to accomplish with this final chapter was to make the reader aware that rodding is made up of cars often very much alike and very much different. Look for the differences in building your own rod while appreciating the differences in others.

A sharp, talented fabricator with an eye for clean lines can often make something so simple it is easily overlooked by the uninitiated. A case in point is the shock mount, headlight mount and directional signal mount all in one. This is Magoo's personal car.

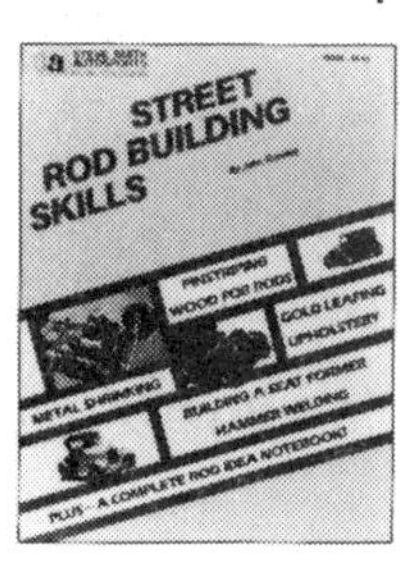

If you want a nit-picking look at all details of rod building, read a companion book from Steve Smith Autosports, "Street Rod Building Skills."

Here's a neat way of constructing an attractive door stop for an early car.

(Left) Rubber bumpers such as this are available from a variety of sources for a number of old cars. They can be most helpful in tightening door fit and eliminating rattles and squeaks.

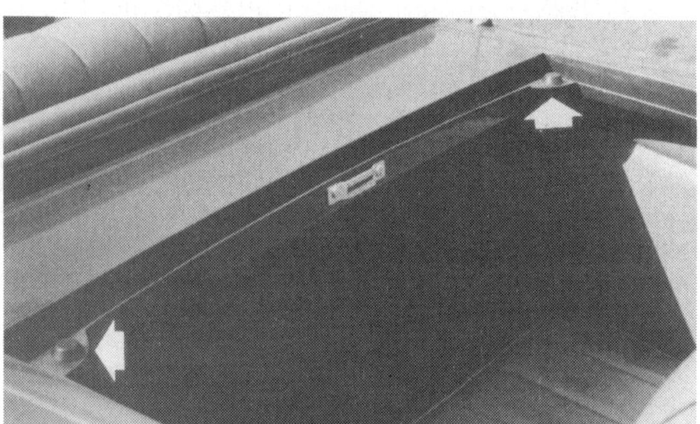

Here's another use for those rubber bumpers. One at each corner of the opening end of a rumble seat or deck lid will facilitate fit.

A positive method of keeping the windshield from going over-center and back into the driver's knuckles can take several forms. This is a fabricated stainless steel clip jammed into the top center of the cowl with the lower bar of the windshield frame pressed against it. Works slick.

There is no end to the amount of work that can go into a rod. The clean appearance of the front of this car is very deceiving. Everthing you can think of (and several things you probably can't think of) has been done — yet it all appears to have "grown" that way to the casual observer.

Brian Brennan's Buick V6 powered Model A highboy features a nice finishing touch in the form of a monster "shower cap" which covers the cockpit area when ol' BB is stuffing his face in a restaurant.

The hood of an early car is a good area which can be customized to reflect the owner's tastes. As you attend rod runs when your car is under construction, make notes, sketches or take photos of various treatments for further study.

(Left) Here's another case where a skilled fabricator integrated shock and headlight mount. Note directional signal light in the side of the painted radiator shell.

Keep in mind that without spending an arm and a leg an engine compartment can be given a clean, functional appearance.

Registration and Insurance

REGISTRATION

As your repro rod is being built you should go out of your way to attend rod runs and meet other rod builders in your area. In your conversations continually ask about the simple, straightforward way of registering a street rod in your state. Each state has it's own procedure. These procedures vary from close-to-impossible to very simple. It is most important you get the complete registration picture set in your mind before approaching the responsible agency in your state. This is very important!

In some states there are now "street rod plates," "historical vehicle plates," "antique plates," and so on. Depending on the circumstance, this can be of benefit or hinderance to the rod builder. If you have problems gaining information about vehicle registration, contact the National Street Rod Association (see Supplier's List) and request the name of their representative for your state. NSRA is a good organization to know about (and join). Not only do they have members in every state; they also have Division Directors and State Representatives. There are nearly 100 volunteer officials within the Street Rod Division of NSRA, and all are street rodders working in your behalf to insure the future of the sport and your enjoyment of it.

NSRA also has something called "Fellow Pages" which is a directory of street rodders who have pledged their assistance to other vehicle owners engaged in the same line of vehicle activity. "Fellow Pages" is an extension of the helping hand attitude which prevails throughout the rodding sport and nearly 5,000 of the association's 24,000 members offer to have their names listed in the booklet each year. Hopefully at this point, you are getting the idea that joining NSRA is a good deal.

INSURANCE

Insurance is not as complicated as registration, but here again you need to get some facts behind you before leaping into a situation you may ultimately regret.

The only type of motor vehicle insurance which states require (when they do) is Public Liability and Property Damage. This is the type of insurance that covers the other guy's car and/or your passengers, but it does not cover the damage you do to your own car. PL and PD insurance is not usually hard to get on a street rod if you have a good driving record, as this type of insurance coverage is regulated by the type of driver you are, regardless of the vehicle you are driving. There are exceptions, but usually you can get PL and PD insurance on your street rod through the same agent that carries the policy on your everyday, late model car.

The type of insurance that protects the losses to your vehicle (such as fire, theft, hail or someone running into you or you into a light pole) is called comprehensive, and this is the type of coverage that is hard to obtain on a special interest vehicle such as a street rod. The thing that makes most insurance companies shy away from these vehicles is the fact that they have been modified. Since they don't have a way to define the word "modified," most companies will just avoid *any* modified vehicle.

Insuring a street rod (modified old car) as an antique or classic is never a good idea as you will have to give the impression that it is a stock car, and when and if a claim is made, you have given the insurance company the loophole it needs to not pay the claim.

When insuring a street rod, make sure the company knows just exactly what you have before you pay the premium. Get it in writing that they are insuring a "modified old car." If you try to hide this from them just to get the insurance policy, you are the only one being fooled, because it's when you need it most — after a loss — that you'll find you really don't have any coverage at all. And, they aren't going to refund the premium money you've given them because *you* misrepresented the vehicle.

Remember, insurance companies want to sell you a policy and they want you to pay the premium — but, they don't want to pay claims. Know what you're paying for and what they think they're insuring before yor accept their coverage.

Do not approach this lightly. What you do not need is a local insurance agent telling you the replacement value of something registered as a 1929 Model A Ford Roadster is worth $8,400 when you have receipts proving you've already spent twice that amount. There are insurance companies around that will essentially let the rod owner set the value of the car (within reason) and then insure it for that amount. As this is written the charge is in the neighborhood of twenty dollars per year per each one thousand dollars of value.

Here again, if you can't get the problem solved, check with NSRA as they have an insurance program available to their members which is very good and workable. This is true "street rod" insurance, and the company handling this for NSRA members has a good understanding of the vehicles and how to insure them.

Suppliers Directory

Ai Fiberglass, Inc.
6599 Washington Blvd.
Elkridge, MD 21227
(301) 796-4382

Al's Antique Auto Parts
9225 Gilardi Road
Newcastle, CA 95658
(916) 633-2179

Alden Shocks
P.O. Box 3758
City of Industry, CA 91744
(213) 333-2465

Antique Ford Parts
1441 N. La Cadena
Colton, CA 92324
(714) 825-2929

Automotive Plastics Co.
Box 408 Pleasant Ave.
Geneva, OH 44041
(216) 466-2153

Bill's Rod & Custom
523 W. Main St.
Springfield, OH 45504
(513) 390-0507

Bodies by Art Gooding
7060 Hazard, Unit E
Westminister, CA 92683
(714) 847-2058

Bradley Brown - Woodcraftsman
7821 Alabama, #14
Canoga Park, CA 91304
(818) 347-5902

Butch's Rod Shop
10 E. Main St.
New Lebannon, OH 45345
(513) 687-3110

California Street Rods
17091 Palmdale
Huntington Beach, CA 92647
(714) 847-4404

Dennis Carpenter Ford Reproductions
P.O. Box 26398
Charlotte, NC 28213
(704) 786-8139

Carrera Shocks
5412 New Peachtree Rd.
Atlanta, GA 30341
(404) 451-8811

Challenger Street Rod Components
918 W. Foothill Blvd.
Azusa, CA 91702
(213) 969-3210

Classic Instruments, Inc.
2120 Maple Terrace
West Linn, OR 97068
(503) 653-5600

Custom Auto Radiator
RD 1, Box 453-A
Jackson, NJ 08527
(201) 364-7474

Dayton Wheel Products
1147A S. Broadway
Dayton, Ohio 45408
(513) 461-1707

Bob Drake Reproductions, Inc.
1819 N.W. Washington Blvd.
Grants Pass, OR 97526
(800) 221-3673

The Deuce Factory
424 West Rowland Ave.
Santa Ana, CA 92707
(714) 546-5596

Don's Commercial Plating
443 S. Cameron St.
Harrisburg, PA 17101
(717) 232-5689

The Early Ford Store
2141 West Main St.
Springfield, OH 45504
(513) 325-2408

Esajian Enterprises
P.O. Box 1203
Fresno, CA 93721
(209) 485-5312

Ron Francis Wire Works
167 Keystone Road
Chester, PA 19013
(215) 485-1937

Ford Parts Obsolete
1320 W. Willow
Long Beach, CA 90810

Gates Products
1495 W. 9th St., Ste. 604
Upland, CA 91786
(714) 946-2186

Kugel Komponents
10821 Whittier Blvd.
Whittier, CA 90606
(213) 695-6935

Magoo's Auto
7630 Alabama, #2
Canoga Park, CA 91304
(818) 340-8640

McFarland Autografix
6822 Forest Haven
San Antonio, TX 78240
(512) 684-2871

NSRA
4030 Park Ave.
Memphis, TN 38111
(901) 452-4030

Pete and Jake's
8827 E. La Tunas Drive
Temple City, CA 91780
(818) 287-6175

Poli Form Industries
334-A Ingalls St.
Santa Cruz, CA 95060
(408) 427-0688

Posies
219 N. Duke Street
Hummelstown, PA 17036
(717) 566-3340

Speedway Motors
P.O. Box 81906
Lincoln, NE 68501
(402) 474-4411

Super Bell Axle Co.
152 M St.
Fresno, CA 93721
(209) 445-1601

T.C.I. Engineering
1416 W. Brooks St.
Ontario, CA 91761
(714) 984-1773

Total Performance Inc.
406 S. Orchard St., Rt. 5
Wallington, CT 06492
(203) 265-5667

Walker Radiator Works
694 Marshall
Memphis, TN 38103
(901) 527-4605